T0202923

Design and Simulation of Rail Vehicles

Ground Vehicle Engineering Series

Series Editor:
Dr. Vladimir V. Vantsevich
Professor and Director
Program of Master of Science in Mechatronic Systems Engineering
Lawrence Technological University, Michigan

Design and Simulation of Rail Vehicles

Maksym Spiryagin
Centre for Railway Engineering, Central Queensland University, Rockhampton, Australia

Colin Cole
Centre for Railway Engineering, Central Queensland University, Rockhampton, Australia

Yan Quan Sun
Centre for Railway Engineering, Central Queensland University, Rockhampton, Australia

Mitchell McClanachan
Centre for Railway Engineering, Central Queensland University, Rockhampton, Australia

Valentyn Spiryagin
Chair of Railway Transport, East Ukrainian National University named after Volodymyr Dal, Lugansk, Ukraine

Tim McSweeney
Centre for Railway Engineering, Central Queensland University, Rockhampton, Australia

CRC Press
Taylor & Francis Group
Boca Raton London New York

CRC Press is an imprint of the
Taylor & Francis Group, an **informa** business

MATLAB® is a trademark of The MathWorks, Inc. and is used with permission. The MathWorks does not warrant the accuracy of the text or exercises in this book. This book's use or discussion of MATLAB® software or related products does not constitute endorsement or sponsorship by The MathWorks of a particular pedagogical approach or particular use of the MATLAB® software.

CRC Press
Taylor & Francis Group
6000 Broken Sound Parkway NW, Suite 300
Boca Raton, FL 33487-2742

First issued in paperback 2017

© 2014 by Taylor & Francis Group, LLC
CRC Press is an imprint of Taylor & Francis Group, an Informa business

No claim to original U.S. Government works

ISBN-13: 978-1-4665-7566-0 (hbk)
ISBN-13: 978-1-138-07370-8 (pbk)

Visit the Taylor & Francis Web site at
http://www.taylorandfrancis.com

and the CRC Press Web site at
http://www.crcpress.com

Contents

Preface

Understanding critical rail vehicle design issues and associated dynamic responses is fundamental to guarantee safe and cost-effective operations of modern railways. With the increasing demands for safer rail vehicles with higher speeds and higher loads, implementing more innovative methods for controlling rail vehicle dynamics requires a better understanding of the factors that affect their dynamic performance. Advanced simulation techniques allow such innovations to be examined in detail and optimised before the costly process of introducing them into the operational vehicle environment on a rail network is contemplated.

Coverage is given to non-powered rail vehicles, the various types of locomotives used to haul freight and passenger trains and the self-powered passenger rolling stock used for public transport in many major cities. This book is intended both as an introductory text for graduate or senior undergraduate students, and as a reference for engineers practising in the field of rail vehicle design, maintenance or modification as well as those undertaking research into performance issues related to these types of vehicles. The information provided progresses from basic concepts and terminology to the detailed explanations and techniques that provide a comprehensive understanding of the subject matter.

The following summarises the content covered in each chapter:

Chapter 1: This chapter provides an introduction to the problems of design and modelling of rail vehicles. The applications of rail vehicle dynamics encompass the rail vehicle manufacturing stages of concept development, detailed design, design evolution and risk analysis, to the train operations and track infrastructure maintenance aspects that impact ride comfort, lateral stability, derailment potential, track life and cost of infrastructure maintenance.

Chapter 2: This chapter provides an introduction to the anatomy of unpowered rail vehicles such as freight wagons and passenger cars. Individual components are described and the basic design processes are discussed. A snapshot of applicable standards and acceptance tests around the world is provided. This chapter concludes with a look at recent advances in rail vehicle design.

Chapter 3: This chapter introduces fundamental knowledge of locomotive design and contains information about different locomotive types. The main components of locomotives are described in detail as well as their elements. In addition, a description of different modern traction systems is included in this chapter. The perspective of likely future developments in locomotive design is also discussed.

Chapter 4: This chapter makes the reader aware of the different types of modelling that are applicable to rail vehicles. This chapter also provides a background in the terminology of relevant modelling techniques.

Chapter 5: This chapter introduces the fundamentals of multibody dynamics, which are required to understand the concepts used in the following chapters. Recent developments in rail vehicle multibody modelling software are described and compared.

Chapter 6: This chapter firstly gives an overview of longitudinal train dynamics, then goes into considerable detail on approaches to modelling longitudinal train dynamics. Concentrated focus is given to wagon connection models and, in particular, the more recent innovations, combining this with alternate train configurations (permanently coupled wagon sets, etc.). The usual information regarding modelling traction and dynamic braking systems, rolling resistance, air resistance, curving resistance, the effect of grades and pneumatic braking is included with coverage given to the more recent innovations of AC traction, higher adhesion locomotives and electronically controlled braking. The interaction of longitudinal train dynamics with lateral/vertical wagon dynamics is given in considerable detail as these are now emerging issues in the stability of long heavy-haul trains. This chapter concludes with a section discussing insights from simulation in relation to design.

Chapter 7: This chapter on rail vehicle dynamics is designed to simulate the dynamic interaction of any rail vehicle with virtually any track. A vehicle and track dynamic system is described by a set of dynamic equilibrium equations for any number of bodies, degrees of freedom and connection elements. Some numerical integration methods are applied to solve the equations. Therefore, the dynamic interactions of rail vehicles and track to predict stability, ride quality, vertical and lateral dynamics, and steady-state and dynamic curving response can be investigated. The detailed non-linear modelling of wheel/rail interaction, plus secondary and primary suspension responses are included.

Chapter 8: An introduction to a co-simulation technique is presented in this chapter. The realisation of co-simulation interfaces between existing multibody codes and Simulink is discussed and is based on an extensive literature review and our own experience. The design concept of the co-simulation interface between Gensys and Simulink is described in detail. An example of the application of Gensys/Simulink co-simulation interface is shown in this chapter.

Chapter 9: This chapter describes some simulation cases with high-level task complexity. The authors introduce real examples in order to find typical solutions. In addition, the question of the development of real-time models is discussed based on a literature review and our own experience in this field.

MATLAB® and Simulink® are registered trademarks of The MathWorks, Inc. For product information, please contact:

The MathWorks, Inc.
3 Apple Hill Drive
Natick, MA 01760-2098 USA
Tel: 508 647 7000
Fax: 508-647-7001
E-mail: info@mathworks.com
Web: www.mathworks.com

Authors

Maksym Spiryagin works as a chief investigator at the Centre for Railway Engineering at Central Queensland University (CQU), Queensland, Australia. His current research interests are rail vehicle dynamics, locomotive traction, mechatronics and real-time and software-enabled control systems. He earned his PhD in the field of railway transport in 2004 at the East Ukrainian National University. His research focused on the rail vehicle design and the development of locomotive traction, real-time models and vehicle mechatronic systems. He has more than 80 scientific publications and 20 patents as one of the inventors.

Colin Cole is the director of the Centre for Railway Engineering at CQU. He is also the research program leader for the Engineering and Safety Program of the Australian Cooperative Research Centre for Rail Innovation. His PhD was in longitudinal train dynamics. His rail industry experience includes track maintenance, rolling stock and vehicle dynamics, simulation and the development of on-board devices. His current research interests are train and wagon dynamics, simulation and train control technologies. He has published 72 papers and 1 book chapter, and has 2 patents.

Yan Quan Sun works as a senior research engineer at the Centre for Railway Engineering at CQU. His current research interests include rail vehicle dynamics, longitudinal train dynamics, rail vehicle–track interaction dynamics, and rail–track and bridge dynamics. He came to Australia in 1998 and earned his PhD in the field of railway transport in 2002 at CQU. He has published more than 70 scientific and academic papers.

Mitchell McClanachan is a mechanical engineer and has been involved in railway research projects for individual railway companies and cooperative rail research agencies at the Centre for Railway Engineering at CQU since 1995. His areas of expertise include train simulation, wagon simulation, rolling stock testing, instrumentation, data acquisition, structural fatigue, energy optimisation, hybrid locomotive systems, economics, human factors, railway safety systems and automated monitoring systems. He has published numerous research reports, consulting reports, journal articles, conference papers, patents and short stories. Mitchell is a registered professional engineer of Queensland, a member of Engineers Australia and a member of the Australasian Association for Engineering Education.

Valentyn Spiryagin earned his PhD in the field of railway transport in 2004 at the Volodymyr Dahl East Ukrainian National University, Lugansk, Ukraine. He is now with the chair of railway transport at the same university. His research activities include rail vehicle dynamics, multibody simulation and control systems. Currently, he works on rail vehicle design and dynamics, mechatronic suspension systems for locomotives, locomotive traction and embedded software development. He has published more than 60 scientific papers and has 28 patents as one of the inventors.

Tim McSweeney has over 30 years of experience in the field of railway infrastructure asset management, specialising particularly in track engineering in the heavy-haul environment. He was the senior infrastructure manager overseeing the Bowen Basin export coal network for Queensland Rail from 1991 until 2001 when he joined the Centre for Railway Engineering at CQU to follow his interest in railway research. He retired in 2007, but has continued his involvement as an adjunct research fellow and was awarded an honorary Master of Engineering degree by CQU in 2011.

1 Introduction

There exist great numbers of different designs of rail vehicles, but the structure of such vehicles commonly has a set of standard modules, units and mechanisms which are, or can be, produced by different manufacturers and have different characteristics and behaviour depending on specific parameters chosen by the designers, but their physical nature is still the same. In this book, we provide a general description of the design of the common features of rail vehicles and show the methods used to simulate and verify them. In most cases, the latter is quite a complex task and not possible to do based only on the theoretical knowledge because the reactions to the variations in operational conditions of such a complex system and its component parts are nonlinear and uncertain. Therefore, knowledge and expertise obtained from experimental studies are essential to producing an optimal rail vehicle design. During the writing of this book, the authors generally used expertise in this field obtained at the Centre for Railway Engineering (CRE) at Central Queensland University. The centre is a rail industry-focussed research organisation established in 1994 at Central Queensland University's Rockhampton campus. During its life, the CRE has performed many research projects for specific industrial partners and for the national rail industry more generally through the Cooperative Research Centre Program of the Commonwealth of Australia. Some results obtained from these latter projects, and especially simulation methodologies used in them, have been drawn upon as the basis of much of this book.

The CRE operates a Heavy Testing Laboratory which has been developed with complete flexibility for carrying out experiments across all research projects in order to get accurate information on the behaviour of different systems for further modelling processes. The laboratory consists of portal frames, jigs, a 'strong floor' with great variability in the location of portal frame hold-down points and sophisticated hydraulic equipment designed for maximum flexibility in testing procedures as shown in Figure 1.1. For rolling suspension testing [1], the suspension characteristics are very important for accurate modelling of vehicle system dynamics, and 4-point hydraulic control has been used. The hydraulic servo actuators used in this research provide up to 2 MN multiaxis static load capability and multiaxis fatigue testing up to 0.5 MN and 10 Hz cycling frequency that allows testing configurations to be designed to diverse specifications.

Hydraulic test equipment is controlled by a CRE-developed control system software, which also allows for maximum flexibility in the control of specimen-testing parameters.

Another good example of such experimental work is the bogie rotation testing with the special test rig shown in Figure 1.2. Some investigation results in this field have been published in [2,3]. During this testing, the following behaviours have been validated:

- Centre-bearing longitudinal movement in transitions due to track twist loads;

FIGURE 1.1   Laboratory environment for rolling stock suspension testing.

FIGURE 1.2   Bogie rotation test rig in the Heavy Test Laboratory at the CRE. (a) End view, (b) side view.

- Change in effective rotational friction resistance due to centre-bearing tilt;
- Change in bogie rotation warp deflection;
- Bogie rotational friction measurements.

Some full-scale laboratory tests shown in Figure 1.3 have been carried out for the evaluation of the effect of braking torque on bogie dynamics [4]. During those tests, theoretical and experimental investigations have been performed in the following areas:

- Measurement of brake shoe forces;
- Measurement of stopping distance;
- Wheelset skid;
- Brake cylinder pressure control;
- Wheel–rail interface friction.

The extensive train test programs with rail industry partners allowed the CRE to develop and to validate a fully longitudinal train simulation for engineering analysis – the Centre for Railway Engineering – Longitudinal Train Simulator (CRE-LTS). The software has the usual train simulation tools plus many improved capabilities to facilitate research:

- No limit on rolling stock types or train marshalling configurations;
- Detailed wagon connection modelling;
- Coupler angle calculation;
- Simulations synchronised with field data of various formats;
- Virtual driver software for automated simulation studies;
- Force Road Environment Percent Occurrence Spectra (REPOS) data output for fatigue studies;
- Energy analysis.

FIGURE 1.3 Fully equipped bogie for study of bogie dynamics during braking mode.

Importantly, the CRE still retains ownership of the program code so that CRE-LTS continues to evolve as rolling stock designs change and new mathematical models are added. This software has found wide application in different research areas [5–10].

The CRE has extensive experience resulting from train test programs to various client specifications having been undertaken where the research emphasis has been focussed on derailment investigations, train dynamics, train driving strategies and minimising the energy use. During such programs, the investigations have been performed on instrumentation development, train testing and data analysis in the following projects:

- Single Wagon Test Program;
- Freight Multi-wagon Train Testing Program;
- Train Dynamics Management Program;
- Energy Benchmarking Tests;
- Diesel Locomotive Energy Monitoring;
- Electric Locomotive Energy Monitoring;
- On-Train Telemetry Testing;
- In-Cabin Device Testing;
- Comprehensive Train Test Program;
- Infrastructure Wagon Test Programs;
- Intelligent Train Monitor Program;
- Bogie Evaluation Tests.

All these projects led to the establishment of a quality research environment and a strong base for further studies. Some of the results of such research innovation and instrumentation activity are shown in Figures 1.4 through 1.7.

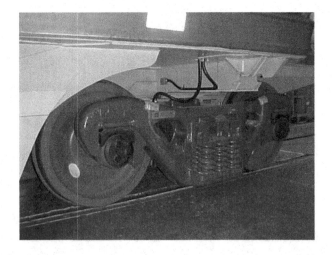

FIGURE 1.4   Bogie strain instrumentation.

FIGURE 1.5 Wheelset-driven generator unit.

In parallel with its testing processes, the CRE is highly committed to commercialising useful research outputs. The following products are available for further development and commercialisation. One of the directions is the work on the design of locomotive bogies. Research evaluated a wide range of passive and active bogie designs using comparative simulations [11–14]. A new active steering bogie design was identified and patented. The design still involves some compromises to ensure the system is adoptable and maintainable. A new active control steering bogie is proposed combining active yaw control of the bogie frame combined with passive forced steering. The new active design, shown in Figure 1.8, can maintain full traction performance up to full adhesion on tight curves.

FIGURE 1.6 Solar cell and telemetry antenna.

FIGURE 1.7   CRE data recorder.

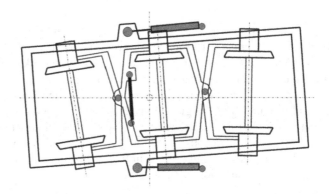

FIGURE 1.8   Active steering bogie developed at the CRE.

In addition, some expertise obtained by two of the authors at the Department of Railway Transport at East Ukrainian National University (Lugansk, Ukraine) allows the inclusion of more information in this book on locomotive design as well as adhesion issues between wheel and rails.

All these examples of previous and current projects show that the team of authors has an outstanding level of expertise in railway research, and the team would like to share this knowledge with readers. In our opinion, the materials presented in this book will be of interest to all technicians, engineers and researchers who are going to undertake their own research in the field of design and simulation for rail vehicles.

REFERENCES

1. C. Cole, M. McClanachan, S. Simson, D. Skerman, Evaluating the performance of 3-piece bogie on short defects and with unequal wheel diameters, *Proceedings of the Conference on Railway Engineering*, Melbourne, Australia, 30 April–3 May 2006, pp. 47–58.

2. S. Simson, B. Brymer, Gauge face contact implications of bogie rotation friction in curving, *Proceedings of the International Conference on Contact Mechanics and Wear of Rail/Wheel Systems*, Brisbane, Australia, 25–27 September 2006, pp. 549–554.
3. O. Emereole, S. Simson, B. Brymer, A parametric study of bogie rotation friction management utilising vehicle dynamic simulation, *Proceedings of the International Conference on Contact Mechanics and Wear of Rail/Wheel Systems*, Brisbane, Australia, 25–27 September 2006, pp. 535–541.
4. M. Dhanasekar, C. Cole, Y. Handoko, Experimental evaluation of the effect of braking torque on bogie dynamics, *International Journal of Heavy Vehicle Systems*, 14(3), 2007, 308–330.
5. Y. Sun, C. Cole, M. Spiryagin, T. Godber, S. Hames, M. Rasul, Longitudinal heavy haul train simulations and energy analysis for typical Australian track routes, *Rail and Rapid Transit*, Prepublished 15 February 2013. DOI: 10.1177/0954409713476225.
6. C. Cole, M. McClanachan, M. Spiryagin, Y.Q. Sun, Wagon instability in long trains, *Vehicle System Dynamics*, 50 (Suppl), 2012, 303–317.
7. Y.Q. Sun, C. Cole, M. Spiryagin, Hybrid locomotive applications for an Australian heavy haul train on a typical track route, *Proceedings of the 10th International Heavy Haul Association Conference*, New Delhi, India, 4–6 February 2013, pp. 751–757.
8. Y.Q. Sun, C. Cole, M. Spiryagin, M. Rasul, T. Godber, S. Hames, Energy usage analysis for Australian heavy haul trains on typical track routes, *Proceedings of the Conference on Railway Engineering*, Brisbane, Australia, 12–14 September 2012, pp. 559–567.
9. Y.Q. Sun, C. Cole, M. Spiryagin, M. Rasul, T. Godber, S. Hames, Energy storage system analysis for heavy haul hybrid locomotives, *Proceedings of Conference on Railway Engineering*, Brisbane, Australia, 12–14 September 2012, pp. 581–589.
10. M. Spiryagin, A. George, Y. Sun, C. Cole, S. Simson, I. Persson, Influence of lateral components of coupler forces on the wheel–rail contact forces for hauling locomotives under traction, *Proceedings of the 13th Mini Conference on Vehicle System Dynamics, Identification and Anomalies*, Budapest, Hungary, 5–7 November 2012.
11. S. Simson, C. Cole, Simulation of traction curving for active yaw—Force steered bogies in locomotives, *Rail and Rapid Transit*, 223(1), 2009, 75–84.
12. S. Simson, C. Cole, Simulation of curving at low speed under high traction for passive steering hauling locomotives, *Vehicle System Dynamics*, 46(12), 2008, 1107–1121.
13. S. Simson, C. Cole, An active steering bogie for heavy haul diesel locomotives, *Proceedings of the Conference on Railway Engineering*, Perth, Australia, 7–10 September, 2008, pp. 481–488.
14. S. Simson, Three axle locomotive bogie steering, simulation of powered curving performance: Passive and active steering bogies, PhD thesis, Central Queensland University, Rockhampton, Queensland, Australia, 2009. See: http://hdl.cqu.edu.au/10018/58747.

2 Unpowered Rail Vehicle Design

2.1 INTRODUCTION

Unpowered rail vehicles are generally referred to as 'cars' in the United States and as 'wagons' in the remainder of the world. In this chapter, all unpowered rail vehicles will be referred to as wagons. Wagons are used to transport various types of cargo and the design varies depending on the type of the cargo, the train consist and rail route. These various design differences will be presented in the following subsections. A wagon consists of a number of components which, depending on the design, can include:

- Body;
- Wagon frame (underframe);
- Bogies;
- Wheelsets;
- Suspension components;
- Couplings;
- Draft gear;
- Brakes.

The locations of some of these typical components on a coal wagon are shown in Figure 2.1. More detail of these components will be provided in the individual subsections.

2.2 TYPES OF WAGONS

The type of wagon is dictated by the cargo or goods it is designed to carry. The two main groups are passenger and freight wagons. Freight wagons include a wide variety of wagon types and designs which are generally named after the cargo or goods they are designed to carry. Freight wagons include heavy haul wagons that carry bulk material, and 'mixed' freight wagons that can carry a variety of freight from construction materials to liquids to retail goods and general freight.

Heavy haul freight wagons are used to carry bulk materials such as coal, iron ore, gravel, grain and sugar; these wagons tend to have high axle loads. Heavy haul freight wagons are typically named after the type of bulk material transported, that is, coal wagons, gravel wagons and grain wagons. Heavy haul freight wagons are also categorised by the design of the wagon. Wagons that discharge their cargo from the bottom are called hopper or 'bottom dump' wagons, while heavy haul freight

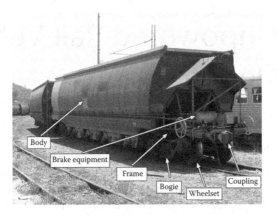

FIGURE 2.1 Typical wagon components.

wagons that require the cargo to be removed from the top, side or via totally inverting the wagon are called bathtub, rotary dump or gondola wagons. An example of a hopper or 'bottom dump' coal wagon is shown in Figure 2.1 and an example of a bathtub coal wagon is shown in Figure 2.2. The construction of heavy haul wagons incorporates the wagon frame within the wagon body.

General or 'mixed' freight wagons are constructed from a frame and the body is built on top of the frame. Again, the wagons are named after the type of freight they carry, from car carriers through to cattle wagons. In the case of freight container wagons, there is only a platform/deck and no wagon body as movable shipping containers are placed directly on top of the frame of the wagon. Figure 2.3 shows two containers on a three-container wagon; these container wagons can also accommodate larger containers by taking up two or three slots. Shipping containers are clamped to the wagon frame. Loading and unloading are carried out by forklifts or gantry cranes. General freight wagons also include wagons that can carry bulk

FIGURE 2.2 Tippler or bathtub gondola-type coal wagon (note rotary coupler).

FIGURE 2.3   Freight container wagon.

FIGURE 2.4   Long haul passenger wagons.

liquids or gases, and these wagons are fitted with longitudinal tanks attached to the top of a frame and are called 'tank' wagons. For liquids of low viscosity, baffles are placed inside the tanks to reduce the dynamic effect of the movement of the liquids.

Passenger wagon design varies for the application of use. Older style long haul-type trains typically use passenger wagons hauled by locomotives at the head of the train, as shown in Figure 2.4. Trains used on shorter city and metropolitan routes are shorter and have power cars at the front and rear of the trains. To create longer trains, the smaller train sets are combined together, causing power cars to be situated throughout the train. This reduces the longitudinal forces placed on the passenger wagons and improves passenger comfort. More modern high-speed trains also have similar arrangements, with power cars distributed throughout the train to be able to obtain the high speeds. An example of a high-speed passenger train is shown in Figure 2.5.

2.3   WAGON FRAMES

The wagon frame (underframe) has two main functions. One function is to support the load the wagon is carrying, and the other function is to transmit longitudinal forces from one wagon to the adjacent wagons in the train. Mixed freight and passenger wagons consist of a low frame on which different body types are attached. The

FIGURE 2.5   High-speed passenger train.

frame in the freight container wagon can be more clearly seen when the containers are removed as shown in Figure 2.6. In this case, the frame of the wagon consists of continuous beams in the centre of the wagon that run the length of the wagon. These support the load of the wagon and also transmit the longitudinal train force to adjacent wagons. Heavy haul wagons, like the coal wagon shown in Figure 2.1, incorporate the frame with the wagon body to allow larger volumetric payloads and, if required, the removal of material through the bottom of the wagon. Longitudinal train forces are transmitted through external beams called 'sills' that run the length of the wagon, and possibly also by a slender beam that runs through down the centre of the wagon. To reduce weight, passenger wagons may also be designed so that the wagon body and frame are incorporated.

Wagon frames are typically constructed of structural steel; this is done to keep the cost of wagons low and because of its ease of manufacture and good fatigue resistance. The wagon frame design process generally uses FEM software to check and refine the design. When designing the wagon frame, it is essential to design the frame to withstand the maximum longitudinal, vertical and lateral forces that will be encountered through normal haulage operations as well as in shunting operations. In long heavy haul trains, the longitudinal forces on the wagons can be very high for the wagons connected directly to the locomotives.

FIGURE 2.6   Freight container wagon frame.

2.4 SUSPENSION ELEMENTS

Rail vehicle suspension is used between the different components of a wagon from the wheels to the wagon frame/body. Suspension elements are termed either 'primary' or 'secondary' depending on the location of the elements. Primary suspension elements are directly connected to the wheels/axles, whereas secondary suspension elements are any suspension element that is not connected directly to the axle. The location of secondary suspension depends on the wagon and bogie design. Wagons may have only primary or secondary or both primary and secondary suspension. The suspension used in typical freight wagons in many countries (Australia, Canada, China, Russia, the United States, etc.) consists mainly of secondary suspension located between the sideframe and the bolster as shown in Figure 2.7. Primary suspension in these types of wagons is generally not used, although in some cases rubber pads placed between the axle bearing and the sideframes function as primary suspension elements. A commonly used freight bogie that has just a secondary suspension is the so-called three-piece bogie. Such bogies are assembled using just three component types: a bolster with sideframes and wheelsets.

In freight wagons, three-piece bogies are used with secondary suspension springs and friction dampers. The friction dampers are based on a wedge design that provides either constant friction damping or, in some bogies, variable friction damping if the spring nests are used to provide the spring force. Constant friction dampers can either have the wedge springs inside or outside the bolster as shown in Figure 2.8. In some bogie designs, the angle of the wedge springs are inclined to the vertical. With variable friction dampers, as the weight of the wagon body increases, so too does the force on the friction wedges resulting in higher friction damping.

Passenger wagons typically do not use three-piece bogies, but use a rigid 'H' bogie with secondary suspension air springs between the bogie and the wagon body. Air springs provide better ride and damping characteristics. An example of a passenger bogie is shown in Figure 2.9.

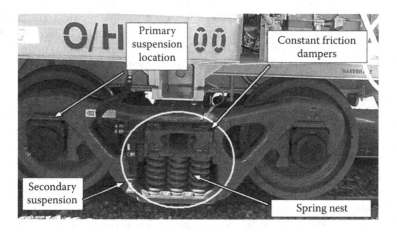

FIGURE 2.7   Typical freight wagon secondary suspension.

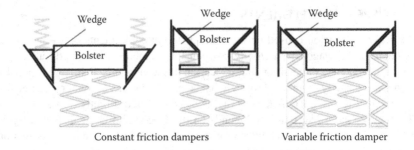

Constant friction dampers   Variable friction damper

FIGURE 2.8   Three-piece bogie friction wedges types.

FIGURE 2.9   Passenger wagon bogie.

2.5   BOGIES

Bogies are the arrangement that contains the wheelsets and are connected to the wagon body/frame. Freight wagons generally use a bogie that is called a three-piece bogie. This bogie contains two wheelsets, two sideframes and a bolster. A freight wagon three-piece bogie is shown in Figure 2.10. The bogie is connected to the wagon body via the 'centre bowl', which allows the bogie to rotate relative to the wagon body. The weight of the wagon provides a frictional damping to rotation. In some cases, a centre bowl liner (a flat polymer disk) is placed in the centre bowl to reduce the rotational friction and allow the bogie to steer more freely. The clearances in the connections between the bolster and the sideframes and the wheelsets and the sideframe allow the bogie to 'warp' or 'parallelogram' when the bogie traverses a curve. Steered three-piece bogie designs have also been developed that use passive linkages to improve the angles between the axles and sideframes when the bogie is traversing a curve. Steered bogie designs reduce wheel and rail wear, but are more expensive to manufacture and maintain. Passenger wagons typically have solid 'H' frame-type bogies with primary and secondary air suspension. These bogies are more expensive to maintain but provide a better ride.

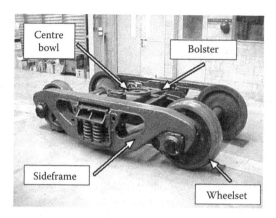

FIGURE 2.10   Three-piece freight bogie.

Bogie components are primarily designed to withstand the static, dynamic and fatigue loading due to the vertical load placed on the components, however small lateral and longitudinal forces do exist due to curving and longitudinal impacts and braking operations, respectively. These additional forces need to be considered when designing the centre bowl and side-frame to bolster connections. The bolster and side-frames in three-piece freight bogies are steel castings to reduce the manufacturing cost; the quality of the material needs to be monitored to ensure the material does not contain inclusions that could initiate fatigue cracks. FEM analysis is used in the design of the components to ensure that they have the desired service life which is typically 20 years. Studies have been conducted to reduce the weight of bogies by using lighter materials and less material, but these designs are generally more expensive and have reduced service lives. As manufacturing quality improves, it is expected that lighter bogies will be adopted.

In freight wagons that use three-piece bogies, there is a tendency for the wagon to rock from side to side on the 'centre-plate' to 'centre bowl' connection. To limit the rocking motion of the wagon body, 'side-bearers' are fitted to the bolster. There are variations in the design of side-bearers, and constant-contact side-bearers as shown in Figure 2.11 are used in the more modern designs. The location of the side-bearers on the top of the bolster can be seen in Figure 2.10. These side-bearers consist of a roller and polymer spring. Constant-contact side-bearers also provide additional resistance to rotation of the bogie, which improves stability but reduces curving performance.

2.6   WHEELSETS AND BEARINGS

A wheelset is an assembly that consists of two wheels fitted to an axle. The tapered profile of the wheel tread allows rail vehicles to remain centred on the track by providing a lateral force on the rail vehicle when it is offset from the track centre. The wheel has a flange to stop the tread moving laterally off the rail. When negotiating tight curves and in severe cases of lateral movement on tangent track, the flange can contact the rail; when this occurs it is called 'flanging'. The interaction of the rail

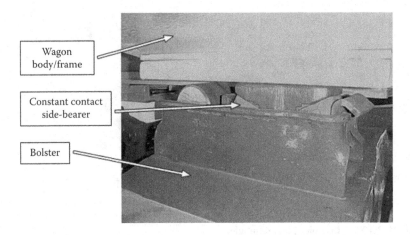

Wagon body/frame

Constant contact side-bearer

Bolster

FIGURE 2.11 Constant contact side-bearer.

profile and the wheel profile is important to the stability and the curving performance of a rail vehicle. The taper of the wheel tread is referred to as its 'conicity'; a high conicity produces larger lateral forces. Figure 2.12 shows a wheel profile; this indicates how slight the angle of conicity of the tread actually is. High conicity is beneficial to the curving performance, but it also lowers the critical speed of the rail vehicle. The critical speed is where the rail vehicle starts 'hunting', that is, moving from side to side due to the high lateral forces caused by the conicity, and the damping in the bogie and wagon is not high enough to damp out this vibration. Above the critical speed, the wagon is said to be unstable. Rail vehicle designers must ensure that the critical speed is higher than the speed of operation.

Wheelset bearings are fitted to the ends of the axles and are housed in a protective assembly. The bearings are typically roller-type bearings as these allow large vehicle loads to be supported. The bearing assembly is connected to the sideframe or bogie

FIGURE 2.12 Wheel profile.

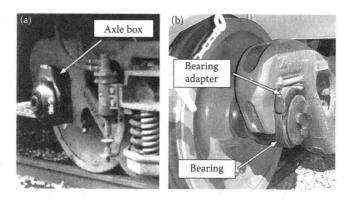

FIGURE 2.13   Bearing types. (a) Axle box, (b) package bearing.

frame either by the primary suspension or a relatively loose connection based on guides with large clearances. The large clearances allow the bogies to 'warp' or 'parallelogram' when the bogie traverses a curve. Earlier bearing designs incorporated the bearing and the guided connection to the sideframe in one component called a bearing box, also commonly known as an axle box. To reduce the manufacturing and maintenance costs, a newer style package bearing is now typically used where bearing adapters are used to securely connect the wheelset to the sideframe or bogie frame. Photos of an axle box bearing and a package bearing are shown in Figure 2.13. It is expected that the weight of the wagon is always downwards on the bearing assembly but, in the case of the package bearings, a bolt and keeper ensure the wheelset remains fixed to the sideframe for extreme instances such as derailments. Bearings have to withstand large transient vertical loads and accelerations due to the wheel-to-rail interface and, usually, the lack of primary suspension between the bearing and the sideframe. Bearing failure is typified by increases in bearing temperatures. Hot bearing detectors are used by some railways to warn of any bearing failure before they can lead to a derailment.

2.7   WAGON BODIES

Heavy haul freight wagons incorporate the frame into the body of the wagon; this is done primarily to decrease the wagon mass and increase the volumetric carrying capacity of the wagon. The material being carried can also dictate the material that can be used for the body of the wagon. Coal wagons typically use stainless steel due to the corrosive nature of coal; aluminium has also been used for coal wagon bodies but stainless steel is preferred as it has better fatigue life and similar strength-to-weight ratio. The body design also depends on the unloading and loading methods used. Heavy haul mineral wagons either unload by discharging the material out of the bottom (typical in coal wagons), or by tipping the wagon on its side or totally inverting the wagon (typical in iron ore wagons). FEM analysis is crucial to the wagon body design process to determine if any stress concentrations exist and to estimate the fatigue life of the wagon under normal operating conditions. The length of the train and the track grades, track condition and payload weight the wagon carries all have to be considered in determining the vertical, lateral and longitudinal forces the wagon could encounter.

Passenger wagon bodies and mixed freight wagon bodies are generally fitted to a wagon frame, although newer passenger wagon designs now use the wagon body as a load-carrying member to reduce the weight. As with the heavy haul wagon designs, FEM is again used to ensure the bodies are capable of withstanding the expected loads and provide a suitable fatigue life. Passenger wagons are also designed for safety (or crashworthiness) in the unfortunate event of a derailment or collision. For this reason, 'anti-climb' devices/features are designed into passenger wagons to stop them from rising up and pushing through adjacent wagons in the event of a derailment. Anti-climb devices are as simple as using beams or columns to strengthen the ends of the wagons to ensure the passenger compartment is not compromised. The analysis would investigate impact performance and the integrity of the passenger compartment.

Additionally, crash energy management (CEM) is increasingly becoming a design technique required by rail transport regulators around the world. This seeks to absorb the energy of a severe frontal collision through placing unoccupied 'crumple zones' at the ends of passenger wagons that plastically deform in a controlled manner. It is crucial that the CEM design also provides for keeping the wagons in-line and prevents over-riding to ensure the transfer of crash energy to more than just the leading wagon. Some rail simulation packages now offer both dynamic and FEM modelling in one package so that crash analysis can be performed on wagon designs.

When designing the wagon body, the allowable size of the rail corridor envelope needs to be considered. This design should account for the maximum expected dynamic movements of the wagon such as wagon body roll and the yaw motion that occurs when curving. The yaw curving calculations use the overall wagon length and the bogie-to-bogie centre distances.

2.8 BRAKE SYSTEMS

The braking system typically used on freight wagons is based on a pneumatic system with manual handbrakes. An air pipe called the 'brake pipe' runs the length of the train which supplies pressurised air to reservoirs on each wagon. Between each wagon, the pipe is connected by a flexible hose called a 'hose bag'. The driver applies the wagon brake by lowering the pressure in the brake pipe; a control valve on each wagon detects this drop in pressure and applies pneumatic–mechanical brakes using the pressurised air stored in the reservoirs on each wagon. A larger reduction in brake pipe pressure results in a larger braking effort. Note that, as brakes are applied by releasing brake pipe pressure, brakes are therefore automatically applied in the event of a hose failure or train pull apart. To apply the braking effort to the wheels, brake shoes are forced against the wheel tread by using pneumatic cylinders called brake cylinders. The maximum force applied is typically dependent on the weight the wagon is carrying; this is done to provide braking without the wheels slipping. Cylinder size and pressures are designed with considerable safety margins to ensure low probability of wheel locking and skidding causing wheel flats. Heavy haul wagons have a load switch that indicates whether the wagon is loaded or not; some freight wagons have variable load valves that vary the brake cylinder pressure based on the height of the spring nest.

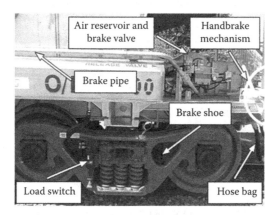

FIGURE 2.14   Heavy haul freight wagon brake equipment.

The wagon brakes are released by re-pressurising the brake pipe; once the brake valve on each wagon detects the increase in pressure, the wagon brakes are released and the wagon air reservoirs are recharged. Some brake valves allow the brakes to be gradually released and other brake valves can only fully release the brakes. Handbrakes are fitted to wagons because the pressurised air can slowly leak out of the wagon brake cylinders causing the brakes to release. The components of the brake system are shown in Figure 2.14.

Modern freight wagons use brake cylinders fitted to the bogie to press the brake shoes against the outside tread of the wheels using the brake beams and guides (see Figure 2.15). There are a variety of brake cylinder arrangements. Some bogies have two brake cylinders while others only use one brake cylinder and a set of linkages to reduce the cost of the brake system. The handbrake uses a ratchet mechanism and is linked to the brake linkages via chains. All brake systems incorporate slack

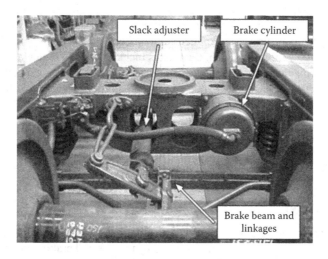

FIGURE 2.15   Freight wagon brake cylinder.

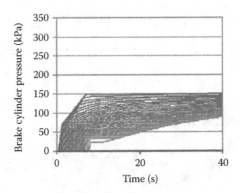

FIGURE 2.16   Air brake delay in a freight train.

adjustment, either in the brake cylinder design or as a separate component called a 'slack adjuster'. Slack adjustment reduces the amount of pressurised air required to apply the brakes.

Because the brakes are applied based on the reduction of air pressure in the brake pipe, there is a delay between the brakes applied at wagons close to the locomotive and those wagons further away from the locomotive. Figure 2.16 shows the brake cylinder pressure in a train with the time delay from the front to the rear of the train being 7 seconds. The delay varies depending on the length of the train and the type of brake valves used on the wagons. Similarly, the release of the air brakes throughout the train is not simultaneous as the brake release is caused by re-pressuring the brake pipe. Delays in wagon braking throughout the train influence the longitudinal train dynamics within the train.

Air brake valves are continually being improved to minimise the response times of braking application and release, particularly in longer trains. However, it is expected that development on traditional air brake valves will reduce as electronic–pneumatic brakes are gradually introduced on rail vehicles. Electronic–pneumatic brakes are actuated via an electronic signal transmitted down the train using an electric cable. Electronic–pneumatic brakes are applied simultaneously on all the wagons throughout the train and allow more control of the brakes including the incremental application, incremental reduction and faster release and re-application. A potential additional feature of electronic–pneumatic brakes is the possibility of automatic handbrakes. This reduces the need for the train crew to apply the necessary number of wagon handbrakes if the train stops on a grade.

Modern passenger wagons typically employ electronic–pneumatic-type brakes with disc brakes. Older style passenger wagons may, however, use the more traditional freight air brakes as described previously.

2.9   COUPLING

Coupling is an important part of wagon design; it affects both the longitudinal and lateral dynamics. Longitudinal dynamics caused by traction and braking inputs and track grade changes are damped out via the damping present in the coupling components.

Lateral forces on wagons are caused by the interaction of longitudinal forces and coupler angles. The longitudinal forces can include steady-state forces due to constant braking/traction levels and grades, or transient forces due to changes in braking/traction levels and grades. The lateral force created by the longitudinal force is dependent on a number of factors including the wagon offset from adjacent wagons, track curvature and the coupler and bogie-spacing dimensions of the wagon and adjacent wagons.

2.9.1 COUPLING MECHANISMS

Modern freight wagon coupling systems use autocouplers. This style of coupler is used in the United States and Australia. Older European freight wagons also employ buffer and drawhook systems, but these are considered inferior to autocouplers due to the higher lateral loads that can occur with this type of coupling and the increased propensity for wheels to climb onto or over the rail. A freight wagon autocoupler is shown in Figure 2.17. These couplers allow limited vertical movement and rotation. The angle of the couplers is restricted either by the design of the coupler housing or 'coupler pocket', or by external guides fitted to the headstock. The knuckle and pin assembly of the autocoupler is the weakest part of the coupling; if large coupler forces are experienced, these components fail first thus reducing the damage to the wagon frame.

Autocouplers have an amount of clearance or 'slack' in the mechanism; this slack accumulates in long trains and can contribute to undesirable longitudinal train dynamics. To reduce the slack in long trains, some heavy haul and freight wagons are permanently coupled where groups of two or more wagons are connected using rigid links or drawbars as shown in Figure 2.18.

Modern passenger wagons also employ slackless couplers as passenger comfort is reduced if too much slack occurs in the couplings. In more modern passenger rolling stock, the inter-wagon couplings have minimal slack and include the pneumatic and electrical connections.

2.9.2 LONGITUDINAL DAMPING

Longitudinal damping is provided by elastomeric elements or friction damping elements or a combination of both. The damping elements are located in either the draft

FIGURE 2.17   Freight wagon autocoupler. (a) Side view, (b) top view.

FIGURE 2.18   Freight wagon solid drawbar.

gear (drawgear) unit or in buffer assemblies on older rolling stock. In an autocoupler design, the draft gear is situated between the coupler and the wagon body using a yoke carrier system as shown in Figure 2.19. This style of mounting allows the draft gear to operate in compression for both tensile and compressive coupler forces. The draft gear has to provide damping for both normal train dynamics as well as the large impact forces encountered during shunting operations. The friction damper provides a relatively stiff connection for the normal low and slowly applied longitudinal train forces; when larger and faster loadings are applied, the friction elements break free, allowing energy to be absorbed as the draft gear moves, thus absorbing the impact energy. This design provides the desirable combination of a relatively stiff connection during normal train operations, limiting longitudinal movements, and a softer connection during shunting operations, absorbing energy during impacts. Draft gear impact characteristics are typically determined by 'drop-hammer' tests where a large mass is dropped onto a vertically mounted draft gear under test. These tests are good indicators of 'shunting' performance and give a good measure of the

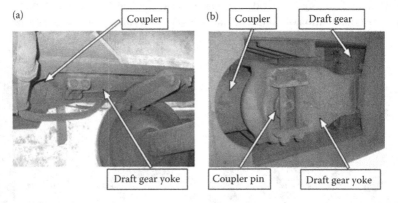

FIGURE 2.19   Location of draft gear under freight wagon. (a) View from side, (b) view from below.

maximum capabilities of the draft gear. Conversely, 'drop-hamper' tests do not show the characteristics of the draft gear for slowly applied in-train forces.

Compared to freight wagons, passenger wagons have much softer longitudinal stiffness and damping due to the lighter weight of the wagons, the shorter trains and the desire to provide a better ride for the passengers. Passenger wagons typically utilise more expensive gas and hydraulic dampers to provide a better and more consistent longitudinal damping for both normal operation, shunting operations and emergency situations.

2.10   STANDARDS

Standards for the design and testing of rail vehicles are governed by different requirements in regions throughout the world. In the United States, the Federal Railroad Administration (FRA) stipulates regulations, while other bodies such as the Federal Transit Authority (FTA), the American Public Transport Association (APTA) and the Association of American Railroads (AAR) provide guidelines. Apart from the individual country standards and regulations, common European organisations that provide universal rail vehicle design standards include the International Union of Railways (UIC) and the International Organisation for Standardization (ISO). In Australia, the Rail Industry Safety and Standards Board (RISSB) are producing a national set of standards/guidelines for the design, maintenance and operation of rail vehicles. RISSB is a relatively new organisation and the national standards being produced are to replace Railways of Australia (ROA) standards, which were observed by State Government-owned rail monopolies prior to 1999. The standards and guidelines for rail vehicles used throughout the world cover the same design aspects irrespective of the region. To provide an introduction to the range of typical rail wagon design standards, the relevant Australia standards that are currently being developed are introduced in the following text. These standards are being developed in conjunction with the current Australian rail industry organisations. These standards aim to maintain, upgrade and harmonise the pre-existing ROA standards and current industry practice. The RISSB standards are published using the Australian Standards designation 'AS'. In relation to wagon design, these guidelines cover:

- Rolling Stock Outlines (AS7507);
- Track Forces & Stresses (AS7508);
- Dynamic Behaviour (AS7509);
- Braking Systems (AS7510);
- Exterior Environment (AS7512);
- Interior Environment (AS7513);
- Wheels (AS7514);
- Axles (AS7515);
- Axle Bearings (AS7516);
- Wheelsets (AS7517);
- Suspension (AS7518);
- Bogie Structural Requirements (AS7519);

- Body Structural Requirements (AS7520);
- Couplers & Drawgear (AS7524).

Where applicable, the standards have separate parts for freight wagons, passenger wagons and locomotives.

2.11   ACCEPTANCE TESTS

Acceptance tests form a part of the railway standards to ensure new or modified rail vehicles comply with the standards of the particular region they are being used in. As for the general wagon design standards, acceptance tests cover the same aspects of wagon performance irrespective of the region. In Australia, the RISSB suite of safety guidelines for rail vehicles includes a Rolling Stock Compliance Certification Standard (AS7501). This standard was developed in consultation with the local railway industry to mandate a minimum set of tests that will ensure all rail vehicles operating in Australia are of an acceptable standard. Static and dynamic test requirements are detailed in some of the standards mentioned in the previous section, particularly AS7507, AS7508 and AS7509. Acceptance tests involve physical tests of the rolling stock either using a prototype wagon or a wagon that represents the wagons being certified. As some of the tests involve testing scenarios which would be difficult, expensive or dangerous to conduct, there is provision for parts of the tests to be conducted using rail vehicle simulation software. For safety reasons, it is usual to successfully complete the static tests before beginning any dynamic testing. Likewise, the dynamic tests should also be sequenced in such a way that certain parameters are successfully examined before proceeding to more challenging tests.

2.11.1   ROLL TEST

The roll test is a dynamic test where a section of track is modified with the maximum allowable track perturbations that the wagon is designed to operate on that would cause a roll motion. The worst-case scenario is where the vertical displacements for the left and right wheels and bogies are out of phase by 180°. Maximum roll would also occur when the wavelength of the perturbations and the speed of operation cause the excitation of the wagon to coincide with the resonance roll frequency. During the roll test, the vertical wheel force of the wagon should not fall below the allowable minimum vertical wheel force (i.e. maximum allowable wheel unloading must not be exceeded). Determination of the wheel force is best done using an instrumented wheel. An alternative to this is to measure the spring deflections during the test. Also during the test, the maximum roll angle experienced should not cause the wagon body to exceed the allowable wagon envelope.

2.11.2   TWIST TEST

This consists of a static test where the individual wheels are jacked up to simulate the wagon negotiating a curve transition. The worst possible case of exiting the transition that includes a localised dip is tested. The forces on the wheels are monitored to ensure that wheel force is not below the required minimum static vertical wheel force.

2.11.3   Bounce Test

The bounce test is a dynamic test where a section of track is modified with the maximum allowable track perturbations that the wagon is designed to operate on that would cause a bounce motion. This involves vertical rail perturbations that occur at the wagon bogie spacing (or axle spacing in the case of two-axle wagons). The maximum bounce amplitude would occur when the wavelength of the perturbations and the speed of operation cause the force inputs to coincide with the resonance bounce frequency of the wagon. The vertical wheel force of the wagon should not fall below the allowable minimum vertical wheel force.

2.11.4   Critical Speed

During testing, the wagon should be operated above its expected operating speed to ensure the critical hunting speed will not be encountered during normal operation. This test involves operating the wagon up to and slightly above its design speed on straight track. A suitably sized lateral perturbation of the rail should be placed in one rail to try to initiate hunting motion. Any oscillatory lateral motion of the bogie or wagon should be damped out for all speeds up to the operating speed plus a suitable margin, usually plus 10% overspeed. Limits on lateral accelerations developed under test conditions are included in AS7509.

2.11.5   Curving Test

The wagon should be tested to ensure the ratio of the lateral and vertical wheel forces (L/V ratio) is lower than the maximum allowable limit for the speeds and curve radii that the wagon will be operated at. Also, for the smallest radius curve expected, the wagon should be checked to ensure it does not exceed the set wagon clearance envelope. Again, the vertical wheel forces should not fall below the allowable minimum.

2.11.6   Wind Overturning

Worst-case scenarios should be considered for wind overloading calculations. This would be where the wagon is empty, with the maximum possible surface area and highest centre of gravity. The amount of superelevation when negotiating curves will also affect the possibility of a wind overturning event occurring. The calculated vertical wheel force of the wagon should not fall below the allowable minimum vertical wheel force for the various expected scenarios.

2.11.7   Combination of Events

The combination of events previously described should be considered to ensure the wagon performs adequately in all possible cases. An example of this may be where the wagon is curving and encountering lateral wind loads. Another example would be when the wagon experiences lateral wind loads when stationary on track with superelevation. Combinations of roll due to track perturbations during curving are another possibility of events combining.

2.12   ADVANCES IN RAIL VEHICLE DESIGN

The expected life of rail vehicles is in the order of 20 years or greater. This means that any advances in rail vehicle design are slow to filter through to the operating rail traffic. New designs and technology can also be difficult to implement if they cause interoperability issues with the existing fleet. An example of this is electronic–pneumatic (ECP) air brakes on freight wagons; an ECP wagon would not be able to operate in the same train as non-ECP wagons. In many cases, it is only possible to adopt new technologies with new fleets or as part of major overhaul/upgrade programs.

Despite the difficulty of introduction, the use of ECP brakes in long freight trains will continue to grow. The benefits include allowing longer trains to operate, providing better brake control and allowing more rapid application and release of the brakes.

Reduction in tare weights of wagons is possible through the refinement of bogie, wagon body and wagon frame designs. The benefit of weight reduction is higher in trains that frequently brake and accelerate such as passenger trains. However, with the introduction of regenerative braking systems, the need for lighter wagons may be negated somewhat. Different materials used in the construction of the wheels, bogies and wagon bodies can also provide weight savings, but it is essential to ensure fatigue lives are still acceptable. Recent advances in aircraft design could translate into lighter passenger wagons, although it is recognised that weight savings in aircraft have much higher returns than what would be expected in railway wagons.

Steered bogies are currently available but generally not utilised; this may be due to the higher purchase and maintenance costs and the fact that some rail operators do not receive enough or any reduction in track access charges for operating this type of bogie. Either changes in access charges or reduction in cost of a steered bogie design may bring a change in this area.

Research is ongoing in the area of wheel–rail profiles to improve curving performance, increase stability, control rolling contact fatigue and reduce wheel wear rates. With advances in computational power, it is possible that improvements can be made in this area.

While improvements may be made in one area, it is possible to also create problems in another. Railway designs, materials and systems have evolved and been optimised over more than 150 years. However, better tracking bogies may lead to rolling contact fatigue issues in both wheels and rails if the wheel–rail contact consistently occurs at one location on the wheel and rail.

Occasionally, quite new and radical concepts emerge. Some recent examples include the Cargo Sprinter, which has a low-powered locomotive at each end. The locomotive can carry up to two containers. A short train of flat wagons can be marshalled between the locomotives. Longer trains are achieved by adding more loco wagon sets. In North America, a concept known as the Iron Highway emerged. This was a low-profile, continuous wagon deck, capable of taking on board regular road semi-trailers. New ideas like this often face the same battle as new technologies and usually must gain acceptance without much usage prior to the first fleet-wide adoption.

3 Design of Locomotives

3.1 HISTORY OF LOCOMOTIVES

The history of rail transport development is directly linked with the advent of locomotives and improvements of their designs and their manufacturing. The first locomotive building process can be dated to 1801, with the construction of a steam road car (called the Camborne road engine) which had been designed by the British inventor, Richard Trevithick. The further transformation of that design was done by him with the assistance of John Steele in 1803–1804, when it was re-designed for usage on rails for the Penydarren Ironworks (Merthyr Tydfil, Wales). This locomotive is considered as the first real locomotive in the world. Historical records show that 10 tons of iron, 70 passengers and 5 wagons were drawn by the locomotive from the ironworks to the Merthyr-Cardiff Canal. However, that locomotive was not as good as many other individual locomotives manufactured by other inventors.

The decisive step in the development of this technology belongs to another British inventor, George Stephenson, who built a locomotive for a mine railway in 1814. That locomotive, named 'Blücher', was capable of drawing a 30 tons load (wagons with coal) up a hill at 6.4 km/h. It was the first successful experience of locomotive manufacturing which used friction forces between smooth flanged wheels and smooth rails for the realisation of the tractive effort. After that, George Stephenson established a company named Robert Stephenson and Company with his son Robert as the managing director. The company built several types of locomotives and the first was called 'Locomotion No 1'. This later became a household name for all traction vehicles running on railways.

In addition, the track gauge selected by George Stephenson of 4 ft 8½ in, the so-called 'Stephenson gauge' or 'standard gauge', quickly became the most common in Western Europe and it is still the standard for most railways around the world.

In the second half of the nineteenth century, design of the steam engine had not fundamentally changed; the basic ways for its improvement were sought in the following directions: more power, increased traction, higher speed and also improvements in energy efficiency and operational performance. The work of many engineers and inventors in different countries resulted in steam locomotives at the last quarter of the nineteenth century being more effective traction vehicles at an appropriate level for the science and technology at that time. The development of industry and commerce made a big contribution to the rapid construction of railways and rail transport in the world. By the end of the nineteenth century, the entire world railway network served by steam locomotives was more than 800,000 km.

At the beginning of the twentieth century, the competitors of steam locomotives were beginning to appear. On railway tracks of mining and metallurgical enterprises it was possible to meet electric traction; on the common railways were locomotives with internal combustion power units, that is, diesel traction or diesel locomotion.

These circumstances were the impulse for the further development of the steam engine, but the technical capabilities of the steam locomotive had almost reached its practical limits and could not compete with the fast-growing electric and diesel traction.

By this time, the world economy had a powerful locomotive manufacturing base. Steam locomotives were built in large industrial plants. In the 1930s–1940s, factories produced powerful steam engines up to 3500 hp, which provided for the intensive workload of railways during World War II as well as in the post-war period.

However, the main drawback of the steam locomotive was its low value of energy conversion efficiency, which is no more than 6–15% during train operation, and this was not consistent with the progress of science and technology in the middle of the twentieth century.

During that period, all industrialised countries began the transition to the new, advanced forms of traction which replaced steam locomotives in train operation with diesel and electric locomotives, as well as the restructure of those locomotive companies that had failed to realise the advantages of diesel and electric locomotives. The companies started changing structures, technologies and the organisation of their production line processes. In the United States, it led to a mass production of diesel locomotives by the end of the 1940s. In Europe and in Asia, which both suffered during World War II and its aftermath, the introduction of diesel and electric traction was delayed.

3.1.1 DIESEL TRACTION AND MANUFACTURING OF DIESEL–ELECTRIC LOCOMOTIVES

Creation of the first diesel traction locomotives began in the 1920s with production of individual samples. A diesel–electric locomotive presented new challenges for manufacture in comparison with a steam locomotive, as it is a much more complex and technically diverse machine. Its component parts are not only large cast and welded metal (frames, body and chassis) for which the production technology is similar to the manufacture of steam locomotive designs, but it also requires powerful diesel engines, compressors, fans, heat exchangers, electric machines and more complex apparatus and devices. Therefore, in contrast to the steam locomotive where almost the entire production cycle could occur in a single plant, for diesel locomotive manufacturing it was necessary and still requires extensive cooperation of many industries, especially the heavy machinery, diesel engine and electrical machinery industries. The large volume of production requires different design and technological processes (assembly, installation, etc.), along with laboratory and field tests. The beginning of the industrial diesel–electric locomotive era was initiated by large enterprises in the United States, Canada and the former USSR in small numbers in the 1930s and then in mass production in the 1950s–1960s. The enterprises facilitated the mass transition of these countries to mainly diesel traction locomotives. The major manufacturers of diesel locomotives at that time were North American and Canadian corporations such as 'General Motors', 'General Electric', 'ALCO' and 'Bombardier'. The maximum volume of production of diesel locomotives in the United States came in the 1950s–1960s and reached about 4,000 units per year. In the former USSR, the locomotive production has been mainly concentrated in large engineering plants located in Lugansk, Kolomna and Bryansk, and in those same years, maximal production reached 3000 units per year. In the countries of Western Europe, as opposed to North America and the former USSR, the main focus was on the electrification of railways, and the production of diesel locomotives was significantly lower with the emphasis being placed on their export as well as on local use of industrial and shunting locomotives with light axle loads. The main concentration of diesel traction has been and is still in railways in industrialised countries such as the United States, Canada, Australia, Russia and China with more than 40,000 locomotives. Currently, the world has more than 1 million km of railway track length served by diesel traction. These are mainly situated in the United States, Canada, Russia, India, China, Brazil, Australia, and South Africa.

Currently, diesel–electric locomotive manufacturers are making significant efforts to improve their products with the introduction of the latest advances in science and technology into this field. The main areas continue to be the following: to increase the power capacity per unit, improve traction and operational performance (maximum realisation of adhesion, improve efficiency and environmental performance, maximise operational safety and improve ergonomics for train crews).

3.1.2 ELECTRICAL TRACTION AND ELECTRICAL LOCOMOTIVES

The development of railway electric traction progressed in parallel with the evolution of electrical systems and the creation of electric machines and devices towards the end of the nineteenth century which were capable of implementing the technical parameters and characteristics necessary for the operation of the railways. The first practical prototypes of electric rolling stock were created in the 1920s. Industrial production of electric locomotives began in the 1930s and there was a progressive improvement of continuous traction power capacity and speed of locomotives until World War II. During the war, the production of electric locomotives was completely suspended. After the war, railway operators in Europe renewed their demand for electric locomotives; this was connected with repairing and extending the areas of electric traction. Electric traction became the main form of propulsion in Europe, mainly due to better energy efficiency and higher available power per traction unit, lower locomotive maintenance costs, more sensitive control and a reduction in the impact on the environment.

Modern electric locomotives can operate using different types of voltages and currents, and this is called multi-system performance (allowing operation on alternating and direct currents and at different voltages). Overhead traction wiring systems are generally used except for underground railways where limited clearances generally result in third rail delivery of electricity for trains. Capacity for freight electric locomotives is up to 10,000 kW per unit. In passenger traffic, it is possible to see a shift from passenger locomotives towards high-speed electric passenger trains, which are currently operating at speeds up to 350 km/h. Countries with traditional diesel traction such as the United States, Canada, China, Russia and others have also started the introduction and establishment of high-speed passenger traffic using electric traction.

Leaders in the development and production of electric locomotives, electric trains, electric urban transport, high-speed electric trains and power equipment were and are the long-established concerns such as 'Siemens' and 'Alstom', as well as big firms such as 'AnsaldoBreda', 'ASEA Brown Boveri' (ABB), 'Bombardier', 'Krauss-Maffei', 'Mitsubishi', 'Kawasaki Heavy Industries', 'Hitachi' and others. Design improvements and modular components and parts for electric rolling stock have required the creation of new technologies in the fields of electrical engineering, aerodynamics, super lightweight and durable materials, control systems and safety performance on the track. Taking those into account, many firms are undertaking parallel manufacturing of rail vehicles with electric and diesel traction. The most promising direction is to create a series of locomotives based on a modular principle where a diesel or electric locomotive would be built on the basis of a standardised vehicle with as many common components as possible combined with as many individually specific 'modules' as necessary for the intended rail operation. This trend can be seen in the latest design developments of companies such as Siemens, Alstom, Bombardier and others.

3.1.3 MAGNETIC LEVITATION LOCOMOTIVES (TRAINS)

Owing to the significant increase in traction power capacities and speeds for wheeled rolling stock, some problems arise with the realisation of the required adhesion coefficient between wheel and rail, and it becomes difficult to meet the safety requirements of stability and acceleration and braking performance. In addition, the dynamic impacts that very high-speed trains impose on track components greatly increase equipment deterioration and maintenance costs. As a consequence, there is a new direction for the propulsion of rail vehicles which implements magnetic levitation technology basic to design solutions for a type of rail vehicle that does not rely on adhesion friction. It was developed by Transrapid in Germany during the 1970s–1980s; that company is now one of the subdivisions of ThyssenKrupp. Operation of the magnetic levitation train is based on replacing conventional wheel–rail contact with magnetic blocks on the vehicles and either ferromagnetic plates or electrical coils on the supporting guide way structure, creating magnetic fields that provide both lift and thrust. Lateral clearance between the vehicles and guide ways is maintained in a similar fashion.

The first magnetic levitation line with a length of 30.5 km of double track and a maximum travel speed of 430 km/h was opened for commercial operations in 2004 by the Chinese government between Pudong Airport and the Shanghai Metro station of Longyang Road. It was set up with the participation of ThyssenKrupp and a subsidiary company of Siemens.

3.2 TRACTION ROLLING STOCK

Railway rolling stock can be divided into two broad groups: powered rolling stock (which provides the motive power for a train) and unpowered rolling stock (which is hauled by powered rail vehicles).

Powered rolling stock can be wheeled or use magnetic levitation.

The traction force for wheeled rolling stock, which moves a train, appears as a result of realisation of tractive efforts by a powered vehicle (rail traction vehicle) as a result of the friction process between its wheels and rails.

For rolling stock which uses magnetic levitation principles, the traction force is created by the magnetic propulsion force provided by a linear induction motor.

Powered rolling stock can be classified using two approaches:

• *By energy sources:* This divides rolling stock into two groups—non-autonomous and autonomous.

Non-autonomous rolling stock is usually provided with energy from a source being outside a powered vehicle. Electric locomotives and electric trains are a good example of non-autonomous rolling stock.

Autonomous rolling stock receives required energy for a motion process from a power plant which is mounted directly inside the vehicle. Steam, diesel and gas turbine locomotives as well as hybrid transport vehicles and diesel trains are this type of rolling stock.

Advantages of non-autonomous traction are a possibility to realise higher power by a rail traction vehicle, significant reduction of effects on the environment during operation of such rail vehicles, and also a possibility of the more efficient use of energy (e.g. regenerative braking in electric locomotives).

However, autonomous rolling stock also has its own advantages such as much lower costs of construction and maintenance of transport infrastructure (absence of a network of electrical supply substations, etc.), and also providing a possibility for working in critical conditions and extraordinary situations (failure of the electrical supply stations and substations, loss of connection in contact conductor wire networks in case of bad weather conditions of icing or hurricanes, etc.).

Currently, some design works on the development of rolling stock with combined energy sources are in process which should allow a rail vehicle to work in both autonomous and non-autonomous modes.

• *By types of use:* This allows division into freight or passenger transportation (or both can be combined in one), shunting operations and powered vehicles for special aims (usually special rail vehicles).

Passenger rolling stock consists of two groups, these being locomotives and motorised carriages (cars), including multiple unit trains. Freight rolling stock consists of the locomotives designed for high tractive effort for hauling large freight and heavy haul trains. As well as individual units, these locomotives are used in groups (parallel control) and used in different train locations (distributed control). These locomotives do not usually have any on-board payload capacity (with the exception of the very recent Cargo Sprinter innovation).

Shunting locomotives (switch engines) perform works in stations related to the forming of trains which they assemble for dispatch or disassemble upon arrival. Usually they do not possess the large power capabilities of main line locomotives and are able to work on track with less axle loading.

The rail traction vehicles for special aims can be classified as vehicles which are able to perform special user functions other than transportation of cargo and passengers. For example, there are such vehicles as maintenance of way vehicles, military use rail vehicles or fire fighting and rescue vehicles and so on.

A powered rail vehicle by itself is a very complex engineering construction, which incorporates many achievements and innovations of modern science and technology. Development, testing and research of such complex machines are not possible without the knowledge and design skills, and an understanding of modelling and calculation of structural components and parts applicable in this field.

Unlike other types of transport vehicles, rolling stock must satisfy a great number of specific requirements for its operation on railways. Further, additional requirements for different railways can vary substantially from each other.

Basic limitations, which have a significant influence on the design of rolling stock, can be defined as follows:

- *Track gauge:* Many different dimensions for gauge have found wide application; for example, the dimensions for commonly used gauges can vary from 1000 to 1676 mm;
- *Loading gauge:* Predefined dimensions (height and width) for rail vehicles, which should allow vehicles to be kept within a specific 'swept envelope' that provides adequate clearance to the surrounding structure outlines (e.g. tunnels, bridges and platforms) in order to ensure safe operation of vehicles through them; the loading gauge for rail vehicles is usually different for different countries and railways which means that it can differ considerably from one railway to another;
- *Axle load:* Limits in loadings on a rail are determined by the track structure, including what kind of rails are in use, the types and spacing of railroad sleepers, the bearing capacity of the track foundation (ballast, sub-ballast and underlying formation), and also by strengths of bridges and other engineering support structures;
- *Types of coupling and absorbing devices (also called draft gear):* Couplers are used for the connection of railway transport vehicles in a train, for the transmission of tractive and brake efforts from powered transport vehicles to unpowered ones (e.g. to the wagons and carriages), for absorption of the shock loadings which occur during motion, stops and also during shunting.

The first coupling devices were hooks and screw couplings, which are still in use for connecting rolling stock in a train on some railways. Such connections need to be made manually, and the process is very slow and also risky from the safety point of view. This is a reason why, at the beginning of twentieth century, such devices began to be replaced by automatic draft gear or automatic coupling devices. There are numerous designs of automatic couplings which are in use on different railways and in different countries. The following basic types of absorbing devices are used in automatic couplings:

- Spring-friction;
- Hydraulic;

- Rubber elements;
- Elastomeric elements (elastomers).

The automatic couplings in use on freight-powered rolling stock are usually equipped with a spring-friction design. For passenger operations, they are usually equipped with a similar design which also incorporates rubber elements. Other types of coupling devices are used in high-speed train operations.

In addition, some vehicles with hooks and screw couplings also have buffers which are installed near to the lateral edges (corners) on the front and rear of railway vehicles. This design limits slack in trains and reduces the shock loadings.

- *Signalling and safety systems:* On the railways of different countries, different standards are used to ensure the safety of operations; rolling stock operators often have issues in regard to operational safety in the case of a necessity to run on tracks of different railways and countries; in particular, the European Union develops special normative documents for standardisation of the different signalling and safety systems;
- *Brake systems:* On railways, various brake systems are used that have found wide application, and these systems can be classified as follows:
  - Pneumatic;
  - Electric;
  - Hydraulic;
  - Mechanical.

All these systems can have various design and structural arrangements.

The main distinguishing design features for rail-powered vehicles and their components are described below.

3.2.1 ELECTRIC LOCOMOTIVES

An electric locomotive is a non-autonomous locomotive, which receives electrical power for its motion from an external electrical supply source.

The general scheme for the electrical power supply system used on electrified railways is presented in Figure 3.1. The electricity from the power plant is transmitted to traction substations over the high-voltage distribution power lines. The substations perform the transformation of the current in accordance with the parameters

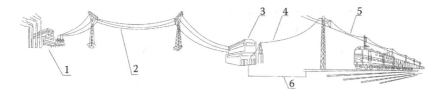

FIGURE 3.1   General scheme for the electrical power supply system of electrified railways. 1—Power station; 2—distribution power lines; 3—electrical traction substation; 4—feeder power line; 5—overhead line equipment; 6—return feeder.

required and then supply it through feeder power lines to points along the overhead line equipment for powering electric locomotives through the contact conductor wire. For closed-loop networks, the railway track is equipped with special return feeders which are connected to the power substations.

Electric locomotives can be divided into three types:

- Direct current (DC) electric locomotives;
- Alternating current (AC) electric locomotives;
- Multi-system electric locomotives.

Electric locomotives can be designed to operate on either DC or AC, or selectively operate on both. Furthermore, the voltage of DC and AC as well as the frequencies of ACs can be different on different railways. The brief classification of electrification systems by currents and voltages commonly used for electric rail traction is summarised in Table 3.1.

Figure 3.2 shows one of the equipment layout options for an AC electric locomotive. DC electric locomotives are different from AC ones because they do not have high-voltage AC electrical power, therefore they do not have a step-down transformer for feeding the DC traction motors.

Multi-system electric locomotives have the current collection, traction and power equipment required for working with several different combinations of current and voltage.

An electric locomotive consists of the following basic systems: electrical, mechanical, pneumatic and hydraulic.

The car body, main frame, coupling devices, suspension, devices for transmission of tractive and brake efforts, bogies and a system for air cooling and ventilation of the electric traction equipment belong to the mechanical system of an electric locomotive.

The pneumatic system includes an air compressor which supplies compressed air through connecting pipelines to the brake system as well as an automatic control system, reservoirs for storage of the compressed air and control and management systems and instrumentation (valves, manometers, etc.).

Contact conductor, power transformers, inverters, traction electric motors, auxiliary machines, electrical control and management units, and the dynamic and

TABLE 3.1

Simplified Classification of Electrification Systems for Railway Networks

Direct Current	Alternating Current	
Voltage (V)	Voltage (kV)	Frequency (Hz)
750	15	16.7
400–2000	25	50
2000–4000	25	60
	12	25

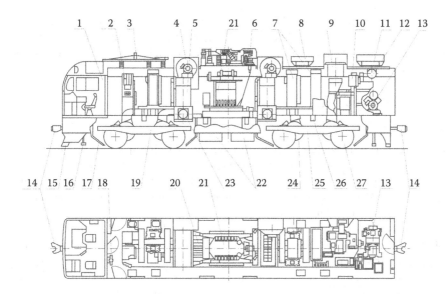

FIGURE 3.2 Example of the main component locations for an AC electric locomotive (Novocherkassk Electric Locomotive Plant, Russia). 1—Driver cab; 2—cabinet with electrical equipment and microprocessor control system; 3—pantograph; 4—front bogie cooling fan system; 5—inverter; 6—high-voltage input equipment; 7—cabinet with electrical equipment; 8—main transformer oil cooling system reservoir; 9—set of brake resistors; 10—propulsion rectifier; 11—brake pneumatic system main reservoir; 12—brake pneumatic system auxiliary reservoir; 13—air compressor; 14—coupler; 15—headstock; 16—locomotive signalling system coil receiver; 17—sand trap; 18—parking brake; 19—front bogie; 20—capacitor block; 21—main transformer unit; 22—batteries; 23—smoothing reactors; 24—traction motor; 25—rear bogie; 26—pivot; 27—wheelset.

regenerative braking systems are all parts of the electrical equipment of electric locomotives.

The hydraulic system includes liquid cooling systems (oil, water, etc.) of electric locomotives, and also a hydraulic control system and instrumentation.

On electric locomotives, the following types of traction motors can be used:

• Brushed DC electric motors;
• AC motors;
• Brushless DC electric motors.

Traction motors are used in the current designs for the dynamic and regenerative brakes with the purpose of reducing wear of the contact parts of the mechanical and hydraulic brake systems, and also for economy of electrical power consumption.

During dynamic braking, the electric energy dissipates as heat from variable resistors; but, in the case of regenerative braking, this energy is fed back into the electrical power contact network, or into on-board storage in the case of hybrid locomotives.

Brushless DC electric and AC motors are the most promising because they produce a large tractive effort and they have smaller dimensions and weight in comparison with brushed DC motors as well as reduced costs for operation, maintenance and repair processes.

3.2.2 ELECTRIC MULTIPLE UNITS

An electrical multiple unit (EMU) is a train used for passenger transportation on city, suburban and regional rail networks and also for high-speed passenger trains.

An EMU comes under the non-autonomous category of rolling stock which receives energy from an external electrical supply source. As with electric locomotives, the EMU traction equipment can be divided into three types: direct current, alternating current and multisystem.

Designs of major equipment and other systems used on EMUs are similar to those of electric locomotives. The difference is that an EMU is a powered train which consists of driving, motor and/or trailer cars in a classic design scheme. The driving car can also be a motor car. In some cases, a power car (similar term to an electric locomotive) can also be added to the configuration of such a train as a separate unit. Trailer cars are usually not used for traction equipment; in rare cases, pantographs and brake air compressor units can be installed on them. EMU trains can have a modular design, often with a shared bogie approach. An EMU train configuration usually includes from 2 up to 16 cars.

Examples of different train configurations with modular design are shown in Figure 3.3. Driving cars have a driver cab from which to control a train. Some additional equipment, storage space and passenger accommodation can be installed in these cars, and they are commonly placed at both the beginning and the end of the train configuration to allow a return journey without having to turn the train. Cars are equipped with motorised or trailer bogies and also with traction equipment and pantographs.

Examples of the typical layouts of the equipment on EMU driving, motor and trailer cars are shown in Figure 3.4. Unlike electric locomotives where the equipment is located in the car body of a locomotive, the equipment on EMUs is installed out of car bodies (under the car frames or on the roofs).

Electric cars can be made in one- and two-level design variants for passenger accommodation (the two-level design is also called a bilevel car design or a double-deck coach design). There also exist other one-level design variants for the increase of passenger capacity by means of placing of seating in two levels.

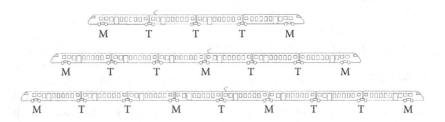

FIGURE 3.3 Examples of EMU train configurations. M—Motorised bogie; T—trailer bogie.

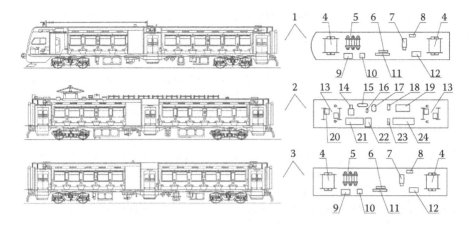

FIGURE 3.4 Example of under frame equipment locations for an EMU (Luganskteplovoz, Ukraine). 1—Driving car; 2—motor car; 3—trailer car; 4—non-motorised bogie; 5—air reservoir; 6, 11, 18 and 23—resistor block; 7—inverter; 8—air brake compressor; 9—batteries; 10, 22—brake equipment; 12, 19, 21 and 24—electric equipment cabinets; 13—motorised bogie; 14—automatic switcher; 15, 16—reservoir; 17—power inductive shunt; 20—traction motor.

For ease of entry and exit for passengers, especially for people with physical limitations, cars with a low-level floor have found wide application. Such a design solution also provides better train stability at high-speed train operation.

The distinguishing feature of trains used in a city service, and which operate with maximum speeds of approximately 100 km/h, is an application of car designs with low floors and equipped with solutions for reducing the noise level from the wheel–rail rolling contact, for example:

- Protective encasement below the underframe using material with sound absorbing properties which prevents the spread of the sound wave from the contact;
- The use of elastic wheels and wheelsets with sound-absorbing properties.

In addition, individually driven wheels have begun to be used in city trains instead of trains with conventional wheelsets.

Suburban trains normally operate at speeds no higher than 180 km/h. They should provide good train dynamics under high rates of acceleration and braking which are associated with the short distances between stations. This is why they have increased numbers of driven wheels or wheelsets in their train configurations. Furthermore, they are not only equipped with standard pneumatic and electric brakes, but can also be equipped with rail brakes and eddy current brakes.

In inter-regional train operations, speeds can reach 400 km/h. Operation at such a speed requires a significant increase in power (e.g. TGV trains develop 12,500 kW), as well as the application of new design solutions to ensure the reliability and safety. These types of trains are widely used with active suspension systems to guarantee tilting in curves, better load transfer between the elements of running gear, levelling the floor, as well as installation of steering bogies and the application of traction control systems for individual driving wheels. To improve the dynamic performance of these trains, it is necessary to reduce the unsprung weight of the running gear. For this purpose, these vehicles are equipped with solid wheels with small diameters up to 600 mm and traction motors with gearboxes hung on the car body. The transfer of torque to the wheelsets is performed by means of drive shafts. Car bodies are manufactured with high usage of light alloy or composite material with fire-resistant properties.

Close attention is given to the aerodynamic design of high-speed trains. This is due to the presence of significant drag forces, as well as a significant increase in aerodynamic noise and vibrations which appear at speeds over 200 km/h and become dominant, exceeding the noise level from the wheels, running gear and traction equipment.

EMUs are one of the primary means of passenger transport and successfully compete in the short and medium distances against road and air transport.

3.2.3 DIESEL LOCOMOTIVES

Diesel locomotives are the most used autonomous rail-powered vehicles. The power plant uses the internal combustion engine, usually running on diesel. Engines which run on petrol (gasoline) are not common on railways due to high maintenance costs.

According to their service operations, they may be divided into the following groups:

- Freight locomotives (in some cases, for trains with large total mass and heavy axle loads, they can be designated as heavy haul locomotives);
- Passenger locomotives;
- Mainline or freight–passenger locomotives;
- Shunting locomotives (also called switchers).

Diesel locomotives usually consist of the power plant and the four basic systems: mechanical, electrical, pneumatic and hydraulic. Main frame (platform) or monocoque car body designs are used for the transmission of tractive and braking efforts generated by a locomotive to other rail vehicles in the train configuration by means of coupling devices installed on them.

The car body of locomotives with diesel–electric transmission is usually divided into the following areas: operator, auxiliary, alternator, engine and radiator modules. As the frame and modules are placed on the bogies, which have some space between them under the middle of the main frame, the fuel tanks and batteries are commonly installed in that space. An example of such a design scheme is shown in Figure 3.5.

The working principle of the locomotive is to convert the energy of the gases produced by combustion processes in engine cylinders into a pressure force on the pistons which is then converted into rotational energy of the crankshaft. This energy

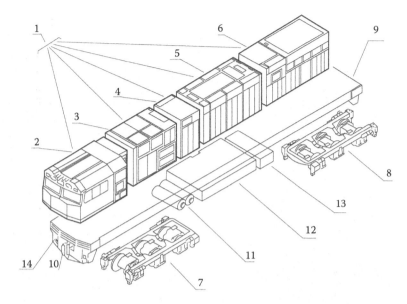

FIGURE 3.5 Example of the main component locations for a diesel–electric locomotive. 1—Car body; 2—driver cab; 3—auxiliary compartment; 4—alternator compartment; 5—diesel engine compartment; 6—radiator compartment; 7 and 8—bogies; 9—main frame; 10—coupler; 11—air reservoirs; 12—fuel tank; 13—batteries; 14—headstock.

is transferred to the transmission system (electric, hydraulic or mechanical) and after that it transforms into the energy for the traction motors, which deliver the traction through a gearbox or directly to the wheels or wheelsets. The traction, which is realised as a tractive force, is needed for the movement of the locomotive and the wagons coupled to it.

Locomotives can be made as one- or two-cab versions, and they can also operate as a multiple unit system (two or more locomotives). An example of a locomotive with two cabs is shown in Figure 3.6.

Classification of locomotives can be based on the characteristics and parameters associated with their basic equipment components installed in the locomotive.

The power plant used in the locomotive can be characterised using the following features—by the number of power plants (one, two, etc.); by the method of the cylinder position (vertical, horizontal, V-shaped, in-line, two-row, etc.); by the type of operating cycle (two-stroke, four-stroke); by the presence of turbo-charged units for the motor power system, the types and number of stages for air cooling systems; speed and performance control systems of diesel engines (electronic, mechanical, hydraulic and composite).

Power capacity, fuel consumption, and temperature of the coolant and oil can be varied over a wide range depending on the type of locomotive and operational conditions.

Diesel engine cooling is most often carried out by means of water or a special fluid that, after being heated by the cooling system of the engine, is itself cooled by

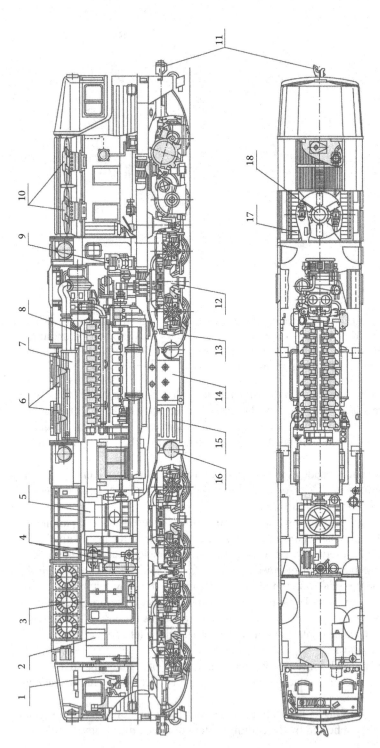

FIGURE 3.6 Example of the general arrangement for a heavy haul locomotive (Luganskteplovoz, Ukraine). 1—Driver cab; 2—high-voltage chamber; 3—motor-fans and electro-dynamic brake resistors, traction motor field weakening resistors, compressor starting and traction generator emergency excitation resistors; 4—air starting compressor and cylinders; 5—centralised air system fan; 6, 10—motor-fans; 7—supercharged air cooler; 8—diesel engine; 9—brake compressor; 11—coupler; 12—bogie; 13—sand trap; 14—fuel tank; 15—battery compartment; 16—main reservoir; 17—cooling sections; 18—water pump.

passing through radiators which are cooled by fans. Diesel oil in older versions of locomotives is similarly cooled, but air-cooled oil is much less effective and is costly in terms of the use of non-ferrous metals. This is why recent and new locomotives have more compact oil–water heat exchangers in which the oil is cooled with water in the cooling sections. In addition, the charge air needed for the diesel engine is also cooled by the diesel engine cooling system. Therefore, most modern locomotives have two or more cooling loop systems in their design. For example, the dual-circuit cooling system of a diesel engine has the primary circuit, where water or coolant are cooling diesel engine parts, and the second circuit for cooling the charge air and hot oil. Advanced and better cooling of the second circuit can increase the reliability and efficiency of the diesel locomotive.

The most widely used diesel locomotive transmission system is the electric power transmission which is characterised by the types of currents used by the main generator (alternator) and the traction motors. They are:

- DC where the generator and traction motors are both DC;
- AC–DC where the generator is AC and traction motors are DC;
- AC where the generator and traction motors are both AC.

Traction motors and generators can be made in brushed, synchronous and asynchronous designs. The traction control system can be an analog, analog-to-digital or digital one.

Locomotive traction motors can transfer a traction torque to just one wheel or wheelset, in which case it is called an individual drive. If one motor is used for more than one wheelset, then it is called a group drive.

The electric transmission system provides optimal tractive and economic characteristics of locomotives.

Mechanical transmission systems are used for locomotives with a low power. Such a transmission is similar to an automotive one, but it has some distinguishing features for the reverse mode of operation.

Hydraulic transmission systems consist of a hydraulic gear box connected to the crankshaft of the diesel engine and mechanical transmission to the wheelsets. The adjustment of traction torque is performed by means of changing the flow rate and pressure of the working liquid (oil). In comparison with the electric transmission, the hydraulic transmission does not need non-ferrous metals and it was widely adopted in the period of electrical copper deficiency during the 1950s and 1960s. However, the hydraulic transmission is a precise machine that requires high-level skills and technical expertise from service personnel, and it also needs high-quality and expensive oils. One more disadvantage of hydraulic transmissions is a lower efficiency compared to electric transmissions.

The auxiliary equipment of the locomotive includes the cooling, air supply and fuel supply systems of the diesel engine, the sanding system, fire protection system, electrical auxiliary equipment and low-voltage circuits and so on.

Descriptions of other systems and components which are part of a diesel locomotive will be provided in the following sections.

3.2.4 DIESEL MULTIPLE UNITS

Diesel multiple units (DMUs) are autonomous multiple unit trains which have diesel engines as their power plant and usually provide passenger transportation in urban, suburban and inter-regional service areas which are non-electrified or partially electrified. In some cases, they can be used as service trains (repair, instrumentation, etc.), and often as cargo transport for companies and factories situated in urban areas or suburbs, providing competition in this sector for shunting locomotives which are frequently used to move specialist cargo or conventional freight cars. The main design elements of equipment on DMUs are similar to diesel locomotives.

Similar to EMUs, DMU trains consist of driving, motor and/or trailer cars in a classic design scheme. The main difference from the electric rolling stock is that, instead of pantographs and electric control circuits of high-voltage equipment, the motor cars have diesel power plants that produce energy transformed then to the traction motors (traction transmission).

Similar transmission types as for diesel locomotives are in use. DMUs can therefore be divided into three categories:

• Diesel–electric (DEMU);
• Diesel–mechanical (DMMU);
• Diesel–hydraulic (DHMU).

The electrical transmission has found much wider application in comparison with others. As for diesel locomotives, hydraulic and mechanical transmissions are generally used with low-powered diesel engines.

There are two common locations for the power plant in DMU trains. The first is a traditional one (see Figure 3.7), where the diesel engine is installed in the driving car behind the driver cab. In this case, this compartment has soundproof insulation on both driver cab and passenger compartment sides. The advantage of such a design is better access to the diesel engine during service or repair works. However, it significantly reduces the size of the passenger compartment.

To increase passenger capacity on modern DMU trains (see Figure 3.8), the diesel unit is often placed in the underfloor space between the bogies of the motor or the driving cars. In this case, the power plants are made up of special packages called modules. If there is a failure, then the relevant module is simply replaced by a new one. The engines in these modules are usually flat engines, where pistons move in a horizontal plane. This is necessary for the reduction of the height of the diesel engines.

One of the trends in the development of diesel multiple unit trains is the application of a configuration where the usually autonomous driving, motor and trailer cars, with minor changes, can potentially be used as non-autonomous (electric) rolling stock on an electrified railway. In such cases, they need to be coupled with a rail-powered vehicle which provides them with the required electric power supply from its own inverter by means of existing electric inter-wagon cable connections. Depending on train configuration, one or more rail-powered vehicles can be used.

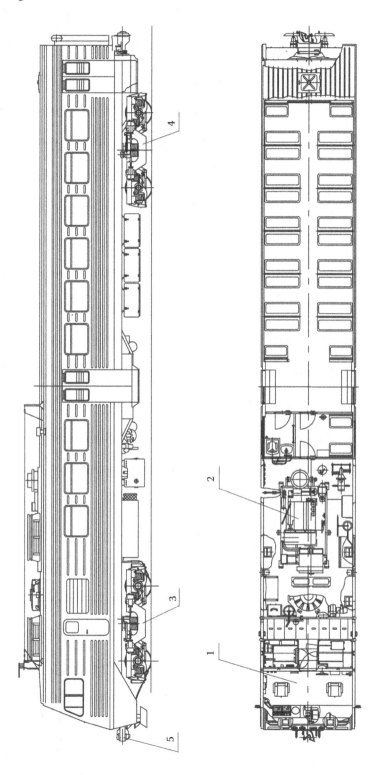

FIGURE 3.7 Example of the classic arrangement for a motorised passenger car (Luganskteplovoz, Ukraine). 1—Driver cab; 2—diesel-generator; 3—motorised bogie; 4—non-motorised bogie; 5—coupler.

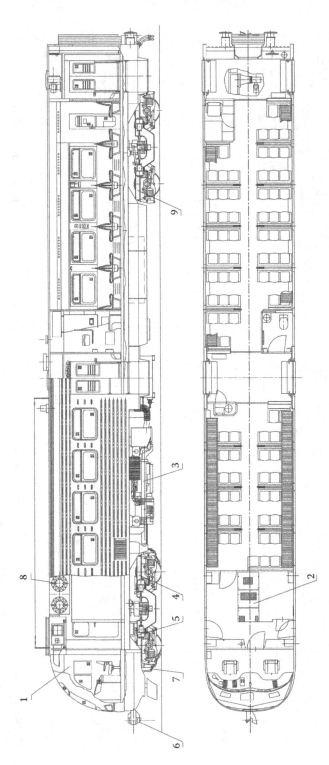

FIGURE 3.8 Example of the underframe power plant location for a motor car (Luganskteplovoz, Ukraine). 1—Driver cab; 2—electric equipment; 3—diesel engine; 4—motorised bogie; 5—traction motor; 6—coupler; 7—sand nozzle; 8—dynamic brake; 9—non-motorised bogie.

3.2.5 Gas Turbine–Electric Locomotives

Gas turbine–electric locomotives are autonomous locomotives which receive the energy required for operation from gas turbines.

Gas turbine–electric locomotives are able to operate passenger as well as freight services. High power capacity and the light weight of power plants also allow the development of high-speed rolling stock traction for non-electrified lines with low axle loads. The layout of equipment on gas turbine–electric locomotives is very similar to diesel locomotives (Figure 3.9).

Gas turbine–electric locomotives are classified by the type of gas turbine plant which may have designs incorporating either a single shaft, or two or more shafts. Currently, the latter types are more used because they can get higher efficiency and capacity extracted from the second and subsequent turbines, providing improved traction characteristics of the locomotive.

For starting the turbines and to allow these locomotives to operate efficiently at less than full load conditions, a small diesel-generator set is often added in their design. This is also often used to reduce the noise when these locomotives operate in communities and places where there are restrictions on noise levels. Furthermore, energy storage batteries, electrical capacitors or pneumatic cylinders with compressed air can be used instead of the auxiliary diesel plant for such operational modes.

Transmission types are identical to ones used on diesel locomotives. Recently, the electrical transmission equipped with AC–AC frequency control design has found wide application.

The main advantages of such a design solution are the high power and simple design as well as the low price of the gas fuel. The main disadvantages are a low coefficient of efficiency which varies significantly in different operational modes, the need for additional equipment for less than full load operation and the high level of aerodynamic noise from the turbine. Taking into account the high fuel consumption, an additional motor car with a fuel tank can be added in the train configuration in order to increase the operational distances for gas turbine locomotives.

3.2.6 Hybrid Locomotives

The high level of competition in the transport market, the tightening of requirements for the protection of the environment and new limits on the consumption of hydrocarbon resources have seen transport engineering communities become substantively engaged in development and introduction of new technologies, including for locomotives, which also includes hybrid locomotives.

By design, these vehicles are similar to diesel and gas turbine locomotives; a significant difference is that, in addition to diesel or gas turbine power plants, hybrid locomotives also use electrical energy stored in electric batteries, supercapacitors or flywheels. The charge process of these components occurs during operation of the diesel generator or gas turbine at idle speed, or when the kinetic energy of braking (of both the train and the locomotive) is transformed into electric power. During hauling operations (traction mode), the combination of energies might be used (i.e. drawing

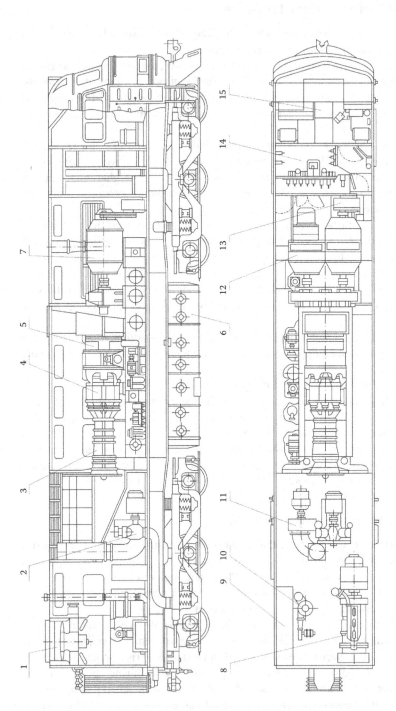

FIGURE 3.9 Equipment locations for a gas turbine–electric locomotive (Kolomensky Zavod, Russia). 1—Cooling compartment; 2—air compressor; 3—turbine compressor; 4—turbine combustion chamber; 5—turbine; 6—heavy fuel tank; 7—traction generator; 8—auxiliary diesel generator plant; 9—diesel fuel tank; 10—boiler-heater; 11—cooling fan for rear bogie traction motors; 12—exciter; 13—cooling fan for front bogie traction motors; 14—high-voltage chamber; 15—driver cab.

simultaneously from the energy storage and the main generator) when additional power is required for acceleration or travelling up long gradients. Further transformation and transmission of energy to the wheels of the locomotive is made in a standard manner as is performed in diesel or turbine–electric locomotives with electric transmission.

The classification of the main systems (mechanical, electrical, hydraulic and pneumatic systems) for hybrid locomotives is similar to that for diesel locomotives. To this classification, the following hybridisation designs can be added:

- Hybrid design with no internal energy storage, only external storage units (hybrid network energy is stored in the energy supply plants or made available to other rail traction vehicles via the overhead line equipment);
- Hybrid construction with internal accumulator units (autonomous hybrid internal energy storage);
- Complex hybrid structures that combine several varieties of these types.

An example of a hybrid locomotive design with internal energy storage is shown in Figure 3.10.

In addition, hybrid traction rolling stock can be divided into groups by the process of regeneration and the energy storage mechanism used:

- Electric when regenerative energy is stored in electrical storage devices such as batteries and super capacitors;
- Hydraulic or pneumatic when energy is converted into internal energy of a liquid or compressed gas or a vacuum;
- Mechanical when the energy is stored in the form of mechanical energy of rotation or translational motion, or its modifications.

Owing to the common application of electric traction transmission in locomotives, the last two of the above-mentioned methods of energy conservation require re-transformation of the stored energy into electricity with inevitable losses.

At the present stage of hybrid traction technology development, hybrid locomotives are already in operation for shunting services, as well as for suburban and urban passenger traffic. However, they are not used for freight or heavy haul traction due to limitations of existing energy storage options. A schematic diagram of the transmission for such locomotives or multiple units is shown in Figure 3.11.

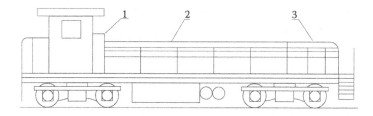

FIGURE 3.10 Hybrid locomotive for switching operations (Railpower Technologies, Canada). 1—Control equipment; 2—batteries; 3—auxiliary power plant.

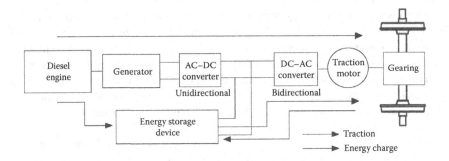

FIGURE 3.11 Locomotive hybrid structure.

For freight or heavy haul hybrid locomotives, large energy storage capacities are required; this is the reason why they would need to use an additional powered wagon (booster) as shown in Figure 3.12.

Batteries are usually used as the electrical energy storage; they can be made from metals (lithium, sodium, nickel, cadmium, zinc, lead) and their compounds, and non-metallic elements (sulphur, carbon, nitrogen, bromine, chlorine) and their chemical compounds.

The main disadvantages are the large weight of the battery cells, their cost, a small number of charge–discharge cycles and a significant change in their characteristics as a function of the ambient temperature of operation, which means the use of hybrid systems equipped with batteries is unpredictable in cold and very hot climates. In those cases, it is necessary to create an advanced system for maintaining the operating temperature of the battery in the predetermined range, which entails an additional energy cost.

Electric drives built on supercapacitors and rotating flywheels (which convert mechanical energy into electrical energy) perform very reasonably and are less susceptible than batteries to the influence of temperature. Supercapacitors can take a charge very quickly and by number of charge–discharge cycles they have a leading

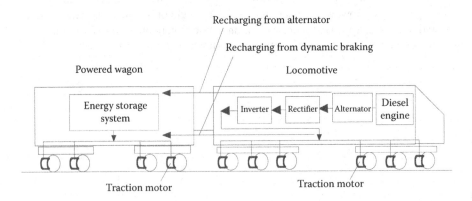

FIGURE 3.12 Conceptual design for hybrid-powered freight train.

position at the current time, but supercapacitors have a shortcoming with regard to the limited time they can hold a charge in storage.

Flywheels have a good ability to store energy and a virtually unlimited number of charge cycles, but are inferior in size and characteristics to both supercapacitors and batteries.

The advantages of the use of hybrid traction are low-energy costs and reduced emissions of air pollutants associated with the power plant in comparison with existing autonomous rolling stock. The disadvantages are associated with the increased cost of hybrid vehicles and the additional operating costs associated with servicing their energy storage systems.

3.2.7 MAGNETIC LEVITATION TRAINS

A magnetic levitation train or locomotive is a non-autonomous train which is levitated on a magnetic field and directed and propelled by means of magnetomotive force arising due to the interaction of magnetic fields between the train and a guide way.

Unlike conventional rail vehicles where traction, braking and guiding efforts, which are dependent on friction conditions, are generated by the vertical load and the resulting adhesion forces between wheels and rails, magnetic levitation trains use the interaction forces between magnetic fields generated on both the vehicle bodies and the guide way to lift the vehicle, set the direction and produce movement of the train. This removes the restrictions imposed by friction forces on the vehicle speed and the implementation of traction and braking forces as for conventional rail rolling stock. The major limitations with magnetic levitation transport are air resistance and the laws of the inductive interaction of magnetic fields of the train and a guide way.

Currently, three common approaches are used in the field of magnetic levitation for trains:

- *Electromagnetic suspension (EMS):* The design principles for this technology are shown in Figure 3.13; EMS is the most energy expending of the magnetic levitation approaches and it uses electromagnets on the train to both lift the vehicle bodies by attraction to a magnetically conductive track and provide guidance of the train along the track.

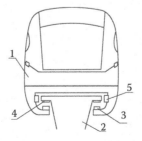

FIGURE 3.13 Magnetic levitation operating on EMS principles. 1—Train; 2—guidance rail; 3—levitation electromagnet; 4—stator (current in track); 5—guidance electromagnet.

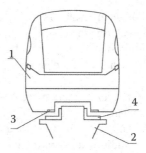

FIGURE 3.14 Magnetic levitation operating on EDS principles. 1—Train; 2—guidance rail; 3—superconducting electromagnet; 4—levitation electromagnet.

- *Electrodynamic suspension (EDS):* This technology provides the train with superconducting electromagnets as well as a guidance rail; the design principles are shown in Figure 3.14. EDS technology enables the development of high speed and is less energy intensive than EMS. At the same time, it requires significant capital expenditure due to the high cost of superconducting materials. Also, unlike the trains using EMS technology, trains which use EDS technology require additional wheels for driving at speeds up to around 100 km/h. When the speed exceeds this value, then these wheels lift off the guide way and the train is flying over the surface of a magnetic guidance rail at a distance of a few centimetres (~10 cm). In the event of an emergency, these wheels (rubber tyres) also allow a softer stop for the train.
- *Permanent magnets (Inductrack):* This technology, shown in Figure 3.15, is similar in design to the EMS approach; Inductrack suspension is the most cost-effective in terms of energy consumption, but the permanent magnets have quite a lot of weight, contain expensive rare earth metals and have a high cost; however, it is simpler from the construction point of view, has less complex control systems, and is one of the most promising directions for the further development of maglev.

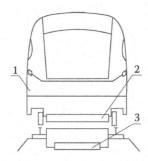

FIGURE 3.15 Magnetic levitation operating on Inductrack principles. 1—Train; 2—permanent magnet on the train; 3—permanent magnet on the track.

Magnetic levitation trains are usually designed for passenger traffic at high speeds and, in this regard, special attention is paid to the aerodynamic characteristics of the train. In order to reduce the weight of car bodies, they are made of light alloys and composite materials using the latest technology.

To isolate passengers from the effects of magnetic fields, they are equipped with special protection devices. In the braking system of magnetic levitation trains, electromagnetic and regenerative brakes are normally used. For emergency braking modes, the trains are also equipped with disc and retractable aerodynamic brakes.

The advantages of this type of transport include low operating costs due to simplicity of the design of the vehicle, low noise impact on the environment, recovery of kinetic energy during braking to be supplied to the vehicle power system and the ability to achieve speeds comparable to air passenger transport. Further advantages include quickly reaching the required speed due to high acceleration and the ability to maintain the speed on rising gradients, which have values significantly higher than acceptable for high-speed rail operations. Limitations of this type of transport are associated with the large energy consumption compared to conventional rail vehicles when driving at low speeds, the limited carrying capacity, the high cost of infrastructure and the complete inability to use the vehicles on a normal railway. Studies have shown that, at high speeds above 300–350 km/h, such transport is competitive with both conventional high-speed rail and aviation passenger transport.

3.3 COMMON LOCOMOTIVE COMPONENTS AND SYSTEMS

In the simulation of traction rail vehicles, it is necessary to have appropriate knowledge and expertise regarding their design, that is, what components and systems they have, as well as a clear understanding of their functions. This section provides an adequate description to represent their main functions in the physical processes for the development of models in multi-body dynamics packages.

3.3.1 CLASSIFICATION OF MAIN COMPONENTS

The powered traction vehicle is a very complex system and can be classified by its components in different methods. However, taking into account that the main idea of this book is design and simulation, this classification should allow an understanding of issues for consideration in the design of traction vehicles and their implementation in the process of the model development. Therefore, in this section, the main components are described from the point of view of rail vehicle dynamists.

3.3.1.1 Locomotive Frames and Bodies

The car body of the vehicle is designed to accommodate its equipment, personnel and, in the case of the presence of a passenger compartment, the passengers, as well as to cope with the application of external and internal loads.

Depending on the structural approach, the car bodies can be divided into two types:

- With a main frame (underframe) as the main load-bearing component;
- Monocoque.

For the first type, all the main loads from the weight of installed equipment as well as traction and braking forces and dynamic and impact loads are received, carried or borne by the strong longitudinal design of the main frame. The side and end walls, the roof and the driver cab are provided solely for the protection of drivers, equipment and passengers from the environment. The first type mainly uses two styles of car body installed on the main frame:

- Cowl unit;
- Hood unit.

In the case of the cowl unit, the locomotive has a full-width car body for the length of the locomotive which is restricted only by the existing loading gauge. The advantage of such a design is the presence of service walkways inside the car body, which allows all service and control works to be done during train operation without leaving the car body, improving working conditions for drivers. If the car body has only one cab, the main disadvantage in this case is low visibility for the driver/s past the other end of the locomotive. Examples of this style can be found in Figures 3.2, 3.6 and 3.9.

The hood unit has side walls and a roof which cover the power and control equipment, and a cab; but, unlike a cowl unit, service walkways are outside of the car body. An example of such a style is shown in Figure 3.5. The main advantage is better visibility in both directions of operation and easy access to the equipment for repair and service jobs.

Car bodies with a main frame have a simple design that allows the assembly and maintenance of locomotives with the body cover elements removed, which reduces the complexity and cost of the work. The disadvantages of this type of car body include a large specific weight, which greatly reduces their competitiveness when creating rolling stock for high-speed or to haul small loads on the rails.

In the case of the second type, a monocoque body has rigid link connections between elements such as the frame, roof and side walls, the tightening belt and so on. It enables collaboration of all elements of the design to resist loads acting on it. This also includes skin elements of the body shell such as the wall-covering sheets. Car bodies of this type are produced in the cowl unit style. The advantage of monocoque construction is the high rigidity and low weight. One of the designs of this type is shown in Figure 3.16.

3.3.1.2 Bogies

Most of the early designs of running gear of powered rail vehicles were without bogies. This was due to the use of crank mechanisms as a traction transmission; these were applied widely in the steam locomotive, and do not allow for displacement or rotation of wheel sets in the horizontal plane. Running gear designed without bogies does not fit well into the curved track sections. To avoid pinching of the running gear with flanged wheels between the rails in curves and turnouts, middle axles of the running gear were equipped with unflanged wheels. Such a design led to extreme wear of the flanges on trailing wheelsets due to the increased guide effort and large angles of attack between wheels and rails. In order to improve the curving

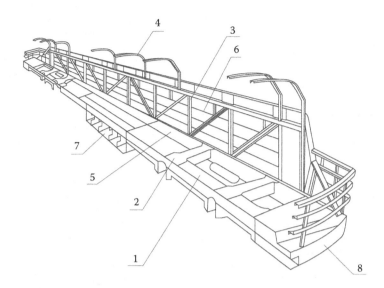

FIGURE 3.16 Locomotive body (Kolomensky Zavod, Russia). 1—Longitudinal sill; 2—cross beam; 3—vertical beam; 4—arc for fixing the roof elements; 5—metal packaging strip; 6—strap for fixation of skin elements; 7—frame for installation of fuel tanks; 8—coupling box.

performance, steam locomotives were beginning to use unpowered carrying wheels and four-wheeled bogies at both ends.

The widespread introduction of running gear with bogies became possible with the introduction of individually driven wheelsets, and this design was defining for the development of diesel and electric locomotives.

The main purpose of the rail vehicle bogie is to improve the dynamic interaction between the running gear and the rails in curved sections of track. In addition, a bogie takes over the support or suspension of the upper weight structure (above the bogie, i.e. the car body) and redistributes it between the wheels or wheelsets through the elastic-damping connection, and also transmits the traction and braking forces to the upper weight structure and coupling devices.

Depending on the design parameters of locomotives (weight, length, tractive effort) and restrictions due to loading gauge and axle load, bogies are available in two-, three- and four-axle design variants. There has recently been seen an interest for bogies with an articulated flexible frame, working on the principle of uniaxial bogies for curved parts of the track and as a bogie with a rigid frame on straight track. A typical three-axle bogie design is shown in Figure 3.17.

The main elements of a bogie are the bogie frame on which are installed the braking system equipment, elements of the locomotive sanding system, spring suspension, wheelsets with associated assemblies and traction drives. Design of locomotive traction drives and their connections with spring suspension and wheels will be described in the next section.

Wheelsets and axle-bearing assemblies (commonly called axle boxes or journal adapters) have a wide variety of designs. Limiting the motion of the wheels in the

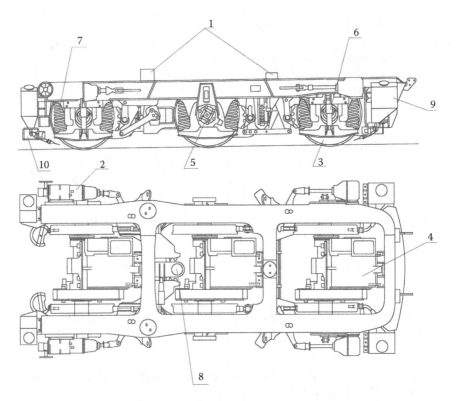

FIGURE 3.17 Three-axle bogie of a heavy haul locomotive (Goninan, Australia). 1—Side bearing; 2—brake cylinder; 3—wheelset; 4—traction motor; 5—axle box; 6—damper; 7—coil spring; 8—yoke for the centre pin; 9—sand box; 10—sand trap.

horizontal and longitudinal planes can be done either by means of elastic connection designs with links (rods) or levers with rubber bushings at connection points, or by means of rigid connection designs with the axle box guide system welded or fixed on a bogie frame.

Axle boxes can be located outside the wheels with the bearing assembly installed at the outer ends of the wheelset, or have an inside location when the assembly is located between the wheels of the wheelset. Inside locations of axle-bearing assemblies are often used on bogies of light rail vehicles and high-speed trains with a low floor design.

Bogies can be designed with radial steering of wheelsets (such a design allows yaw rotation of wheelsets). Bogies also can be equipped with devices for load transfer between wheelsets in the form of various actuators.

To improve tractive force performance on curves, different ways of articulation between the bogies themselves and their connections with a monocoque body or main frame can be used. In addition, wheelsets with independent rotation of the wheels can be used.

3.3.1.3 Locomotive Traction Drives

Traction drives are made up of mechanisms and units engaged in the transfer of kinematic power from the traction motors (electric, hydraulic) or the output shaft of the mechanical gear transmission to the wheelsets or wheels of the powered rail vehicle. Designs of drives are varied and depend on the type and operational service parameters of rail traction vehicles, the selected mode of transmission, the design of wheelsets/wheels and the mounting methods of the traction motor. The traction drive designs can be divided into two types: individual or grouped.

For the individual drive design, the traction torque from the motor acts on one wheelset or one wheel. An example of such a design is shown in Figure 3.17.

For the grouped drive design, the traction torque from the motor or an output shaft of transmission is shared between multiple wheelsets or bogie wheels. The monomotor bogie, which has a grouped drive design, is shown in Figure 3.18.

The design and parameters of traction drives are often dependent on the installation designs of traction motors and associated gearing. Three design variants have found wide application:

- With a nose-suspended traction motor;
- With a frame-mounted traction motor;
- With a body-mounted traction motor.

Generally, the first of these design variants (see Figure 3.17) has traction drives, of which one part is resting on the axle of the wheelset through rolling or slip bearings,

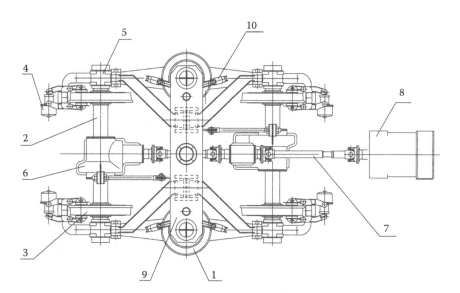

FIGURE 3.18 Monomotor bogie (Luganskteplovoz, Ukraine). 1 —Air spring; 2—axle; 3—wheel; 4—brake cylinder; 5—axle box; 6—gear box; 7—shaft; 8—body-mounted traction motor; 9—bolster; 10—damper.

and the other part is connected through the elastic-damping suspension to the frame of a bogie or the locomotive. Torque from the motor is transmitted to the gear box, the driven gear of which is seated firmly on the axle. The advantage of this drive design is a low price and simplicity of design. It enables the effective transfer of high tractive effort. However, in this case about 60% of the weight of the engine and the traction gear account for unsprung mass; this causes increased dynamic effects of the traction vehicle on the track. This type of suspension is widely used in locomotives with a relatively low design speed, usually on freight and shunting locomotives. This design also has some potential modifications whereby the traction motor rests on and transmits traction torque to the wheelset via elastic elements. The modified design is a bit more complicated, but it leads to a significant reduction of dynamic impact loads which allows its use at higher speeds of up to 200 km/h.

The other two design variants are similar because the traction motor is mounted to the bogie frame or the main frame (car body). An example of a traction drive design with a frame-mounted traction motor is shown in Figure 3.19. Another example with a body-mounted traction motor is presented in Figure 3.18.

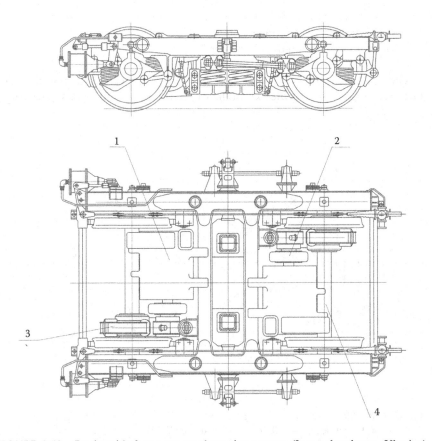

FIGURE 3.19 Bogie with frame-mounted traction motors (Luganskteplovoz, Ukraine). 1—Traction motor; 2—flexible coupling; 3—gear box; 4—wheelset.

In both cases, the wheelset receives a torque through mobile and flexible connection elements that provide the necessary freedom of movement of the wheelset or the wheels relative to the traction motor. In this case, unsprung weight is sharply reduced and this improves the dynamic performance of powered rail vehicles. This type of design is also used on high-speed vehicles.

The wheels can have their own traction drives, providing independent rotation of each of them. In this case, a differential gear is typically used either when both wheels are driven by a single motor, or to drive each wheel using its own motor controlled by the principles of differential gearing with the harmonisation of the frequency of rotation of each wheel.

Depending on the type of traction gear, the drive can be made with an axial gear in which the driving shaft is perpendicular to the axis of rotation of the wheel, or with a radial gear when the axes of the input shaft and the wheelset are parallel.

3.3.2   SUSPENSION AND ITS ELEMENTS

The spring suspension is necessary for a rail vehicle to reduce its force interaction with the track, which arises from rolling contact on track irregularities, and minimise and damp the dynamic forces and the natural oscillations of the vehicle in order to reduce their effect on cargo or to provide passengers with a comfortable ride.

Suspension of a rail vehicle can be performed in several stages (one, two or more), and it acts in the horizontal, vertical and transverse planes.

The primary suspension acts in the vertical plane and it is usually located at the connection points of the wheelset or its axle boxes with a bogie frame or body, but it can also be located inside the wheelset or a wheel (the so-called elastic wheel).

The secondary suspension is commonly located at the connection points between a bogie frame and the car body, but it may also be incorporated between the elements of the bogie itself.

Suspension systems of rail vehicles can include the following elements:

* Elastic elements have stiffness and their task is to allow reciprocal movement of elements of running gear under the force load and under oscillations arising therefrom; these elements have load characteristics;
* Damping elements and shock-absorbing devices have damping properties and are used to absorb vibration energy and reciprocal movement of elements of running gear;
* Elastic-damping elements have combined properties of the elements as mentioned above.

The main characteristics of the suspension system are the deflection and damping values for each of the stages and planes:

* Displacements of elements:
  * Static displacement under the action of the static weight of the vehicle;
  * Maximum displacement, which is limited by the maximum mutual displacement of suspension elements and by the need to remain clear of the

structure gauge (which allows an estimation of the clearance outline) under static or dynamic loading conditions.
• Damping coefficients for each of the stages which show the rate of damping of oscillations of the elements of running gear.

If the set of elastic-damping elements are connected to each axle of a bogie individually, such a suspension is called an individual suspension. If the sets of elastic-damping elements are grouped together using levers and balance beams, the suspension is said to be a balanced suspension. To ensure uniform redistribution of loads between axles and a locomotive's wheels, a combination of the elastic elements in groups is widely used. In this case, one group can be considered a point of suspension. Therefore, it is possible to add one more characteristic to suspension classification, this being the number of 'points of suspension'.

3.3.2.1 Leaf Spring Suspension

Leaf springs are one of the common elements of the suspension of rail vehicles and have both stiffness and damping properties. Stiffness characteristics of leaf springs provide resistance forces from metal leaves, which are part of such a spring, and its flexibility is dependent on the number and thickness of the leaves and their length. All leaves in the spring are covered with spring clamps which limit the relative movement between the leaves in the transverse direction. Disadvantages of leaf springs are their large specific gravity in comparison with other elastic elements, complicated manufacturing process and their poor repairability as well as inconsistent damping characteristics due to the change of the friction force between the leaves.

3.3.2.2 Helical (or Coil) Spring Suspension

Currently, suspension on helical springs, also known as coil springs, has found wide application due to their light weight and their ability to work as a vertical spring, and also act in the transverse plane.

An example of the usage of such a spring in the primary suspension for a locomotive is shown in Figure 3.20.

The property of springs to act in the transverse plane is often used in the second stage spring suspension. Such suspension is also called 'flexi-coil suspension'. An example of such a suspension design is shown in Figure 3.21.

To increase the stiffness of helical springs, they can be combined into sets. Getting non-linear stiffness characteristics is also possible through the use of steel wire with variable wire cross-section diameters along the length, as well as varying the diameter and shape of the spring.

Suspension with torsion springs works on the principle of deformation by torsion or twisting. The advantage of torsion springs is that they have a small mass which is substantially lower than the equivalent weight of the coil spring. However, the manufacturing cost is high for such springs.

3.3.2.3 Air Spring Suspension

Air (pneumatic) spring suspension at the current time is one of the sought-after elements of suspension systems for high-speed and passenger rolling stock due to its

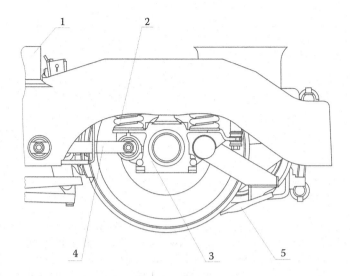

FIGURE 3.20 Primary suspension design of heavy haul locomotive (EMD GM, USA). 1—Side bearing; 2—primary suspension coil spring; 3—axle box; 4—traction rod; 5—sand nozzle.

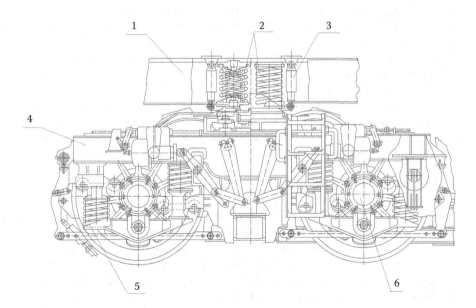

FIGURE 3.21 Suspension design with flexi-coils of a heavy haul locomotive (Luganskteplovoz, Ukraine). 1—Underframe; 2—flexicoil spring; 3—damper; 4—bogie frame; 5—primary suspension coil spring; 6—axle box.

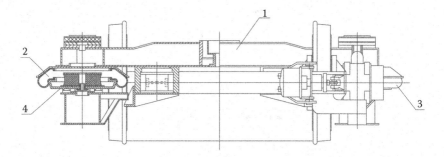

FIGURE 3.22 Example of installation of air spring. 1—Bolster; 2, 3—air springs; 4—rubber emergency suspension pack.

elastic characteristics, which can be adjusted under certain loads and operating conditions, and the ability for load transfer between the elements of the running gear. An additional elastic element is installed inside the air spring in order to prevent suspension collapse during emergency situations and failures. This allows the vehicle to reach a place of repair in cases of failure of the rubber-shell or air feed line.

A typical air suspension system consists of the following elements: air springs, connecting pipes, the levelling valve, an additional reservoir, differential pressure and safety valve.

Changes of the stiffness characteristics are performed by the adjustment of air pressure and temperature parameters. Damping characteristics can be changed by the adjustment of the size of the additional reservoir and the flow area of the pressure valve. Typically, air suspension is used in the secondary suspension because it is more effective in absorbing low frequency oscillations. An example of such an application is shown in Figure 3.22.

Air suspension can have several air springs connected in the loop and several additional air reservoirs. Also, air springs can operate in pairs without the application of an additional reservoir.

The advantages of air suspension are the possibility of varying the stiffness and damping characteristics as well as low weight. The disadvantages are the additional energy costs for feeding air to them and cleaning of the air, and more expensive maintenance and increased cost in comparison with coil and leaf springs.

3.3.2.4 Hydraulic Suspension

Hydraulic suspension works on the principle of a mechanical balanced suspension, but, as with air suspension, it can be divided into circuits which allow different options in organising the damping of vibration. Special oils and liquids for hydraulic transmission and also those commonly used in the hydraulic brake systems have found wide applications in hydraulic suspension systems. The main elements of hydraulic suspension include the following: hydraulic working cylinders, connecting pipes and the master cylinder. The latter has a piston that is connected with an elastic element (coil spring, air spring or torsion bar). This allows adjustment of the required stiffness characteristics of the hydraulic suspension. For the implementation of damping, an additional adjusting system is present which has differential valves.

It also has an additional set of valves and pumps for load distribution. The organisation of individual hydraulic suspension with double-acting hydraulic cylinders is also possible in the rail vehicles. Hydraulic suspension is commonly used for small rail traction vehicles which transfer passengers. The main disadvantage of such systems is the need for high-precision manufacturing solutions for working cylinders and the application of expensive fluids, which in the case of a leak may heavily pollute the environment. Therefore, it has a high cost of operational service. However, this type of suspension ensures good dynamic ride quality.

3.3.2.5 Electro-Mechanical Suspension

For such suspension systems, it is necessary to use magnetic bearings, which are usually classified by the type of magnetic fields used: permanent magnets, electromagnets and mixed.

As the design utilises an electromagnet, solenoid coils can be used because they have adjustable parameters such as supplied voltages and currents. To control them, the microprocessor system reads data from sensors of the car body and bogies' position relative to the heads of rails, as well as from speedometers and accelerometers.

To reduce the energy consumption, such systems can use magnetic fields for damping vibration of elements of the running gear. In this case, the other part of the suspension is designed with standard elastic elements used. Such a scheme is convenient because electromagnetic dampers act as a generator, and this allows use of the energy of vibrations as a power input for the vehicle's own needs.

The disadvantages of these suspension systems are the high magnetic radiation which requires protection consisting of special shielding materials and fittings. In addition, it requires a complex control system that ensures operation of the electro-mechanical suspension in real time. Among the advantages is the possibility of energy recovery as mentioned above and wide ranges of variation of elastic-damping properties.

3.3.2.6 Dampers

In rail transport, the types of damping and absorbing devices are classified by the type of working fluid used in them or by the physical process that creates an absorbing effort.

Dry friction dampers may be designed with a translational characteristic in which a damping force is generated due to a friction process between the piston and the cylinder (their mutual displacement) or a torsional type where the damping force is created by the friction between two or more discs, where one has a rotational motion associated with a torsion arm actuated by motion of movable elements of the running gear. To ensure the constancy of the friction process, compensation has to be made for wear; special mechanisms are used which usually consist of spring elements and tensioners.

These types of dampers are sometimes used in the primary suspension, but can give problems due to the inconsistency of their characteristics and the initial force for displacement, which can lead to locking of spring suspension. During servicing of such rail vehicles, it is necessary to monitor the tightness of the friction elements. The advantages are the simplicity of design and low cost of manufacturing.

Hydraulic dampers (shock absorbers) for damping and absorbing of vibration use the viscous properties of liquids. Usually, they consist of a cylinder in which is inserted a rod with a piston that has drilled holes in it. This makes it possible for fluid to flow from one chamber of the cylinder to another. Flow can also be carried out through channels in the cylinder walls. These dampers have stable damping characteristics for low-frequency vibrations, but they are very sensitive to high frequency because the latter is associated with liquid cavitation processes and hydraulic impact. The performance of these dampers is significantly affected by the ambient temperature and the temperature of their fluid. Often, this type of damper is installed in the secondary suspension.

Gas shock absorbers are filled with gas under high pressure and work on the same principle as hydraulic dampers. However, they do not have the disadvantages associated with the liquid flow process, and can therefore be used in the primary suspension.

Rubber dampers or rubber-absorbing elements act based on the damping properties of rubber. To increase the stiffness and strength characteristics of the rubber elements, they are covered and reinforced with metal or composite materials, fabrics and fibres. They can be used in primary and secondary suspensions.

Combined dampers integrate several types of dampers listed above. Among them, for example, are gas–hydraulic dampers that find wide application. Such dampers are also used in the primary suspension.

3.3.2.7 Combinations of Several Suspension Elements

Usually the suspension systems of the running gear of rail vehicles use different combinations of elastic and damping elements in order to obtain non-linear characteristics or required stiffness and damping properties. Very often these combinations are used in a balanced spring suspension. For example, the combination of coil and leaf springs was used in the primary suspension of the first diesel locomotives.

3.3.2.8 Active Suspension

The development of passive suspension systems is currently approaching its practical optimisation limit. The needs for further security and stability of operation with higher traction and braking forces require designers to create suspension systems for traction rolling stock that could provide opportunities for the redistribution of loads at different operational conditions, reduction of the centrifugal forces, change of frequency ranges of vibrations and the possibility of utilisation of the energy from oscillations. These can all be made possible by introducing a complex system that controls processes in the suspension systems of running rail vehicles. Suspension, equipped with a control system, is called active suspension. The main elements of the active suspension are conventional springs and dampers, and special devices that generate control efforts (actuators), which can be integrated or concatenated with other suspension elements, sensors and a control system.

Active suspension systems can be classified by their main functions:

- Active damping;
- Active steering;

- Active tilting in curves;
- Load transfer between wheels (or wheelsets) and bogies.

An example of a design with active tilting suspension is shown in Figure 3.23.
In addition, the active suspension systems can also be classified by types of actuators:

- Hydraulic;
- Mechanical;
- Pneumatic;
- Electrical and electro-dynamic;
- Magnetic and magneto-dynamic;
- Complex or combined.

Active suspension systems allow the achievement of better characteristics and dynamic results for rail vehicles. However, to ensure that they work properly, it is necessary to have highly qualified personnel servicing such systems and it also results in increased energy costs for the activation of actuators. Consequently, the economic effect of the introduction of these systems may not always provide a significant result, especially if the vehicle is operated at low speeds and with sufficient adhesion coefficients.

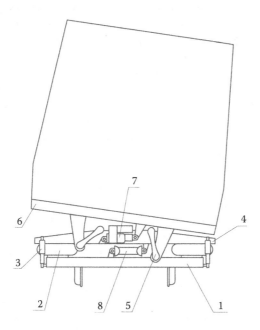

FIGURE 3.23 Example of active tilting suspension system (Siemens, Germany). 1—Bogie frame; 2—air spring; 3—damper; 4—pendulum beam; 5—tilting link; 6—car body frame; 7—tilting actuator; 8—lateral stability actuator.

3.3.3 Connection between a Locomotive Frame and Bogies

Connection elements between a locomotive frame and bogies are used for supporting the locomotive car body on the bogie frames, the transmission of traction and braking forces from the bogies to the car body, and can also be parts of the secondary suspension. Such elements make possible rotations and displacements of bogies relative to the car body within the prescribed limits and their return to the initial position. These include: pivot assemblies, side bearings, links and linkages, return devices and flexi-coil suspension. In order to transmit traction and braking forces between them, traction rods are also used.

An example of the installation of connection elements on the bogie frame of a diesel–electric locomotive is shown in Figure 3.24. The next subsection provides a more detailed description of the basic design of connection elements.

3.3.3.1 Centre Pivots

Pivot assemblies are used to transmit traction and braking forces from the bogie to the car body or the main frame of the locomotive. The pivot assembly is also the point about which a bogie undergoes rotational movement in a horizontal plane relative to the car body.

Pivot assemblies can be divided into two types which are characterised by their position relative to the centre of wheelset axles or wheels in the horizontal plane:

- *With the high location of the pivot point:* In this case, the force is transmitted from the bogie to the car body at a point located higher than the centre of the wheelset in the horizontal plane;

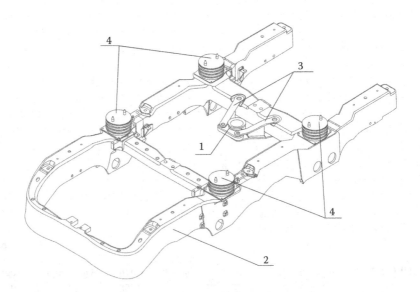

FIGURE 3.24 Connection elements mounted on the bogie frame (EMD GM, USA). 1—Yoke; 2—frame; 3—traction rod; 4—side bearing.

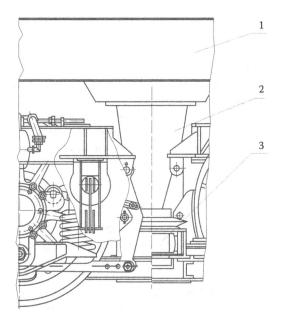

FIGURE 3.25 Low-positioned centre pivot assembly (Luganskteplovoz, Ukraine). 1—Underframe; 2—centre pin; 3—low-positioned centre pin connection assembly.

- *With the low location of the pivot point:* In this case, the force is transmitted from the bogie to the car body at a point located below the centre of the wheelset in the horizontal plane. An example of such a design is shown in Figure 3.25.

When these points have low locations, then a higher value of tractive and brake efforts can be achieved by a locomotive in comparison with a locomotive which has the same design and configuration, but has pivot assemblies with high pivot points.

Pivot assemblies of locomotives can have a rigid design when the bogie can perform a translational motion in the vertical plane and a rotation in the horizontal plane. In addition, pivot assemblies can be designed with additional gaps, which allow some small motion in the horizontal plane transverse to the longitudinal axis of the locomotive.

Pivot assemblies with spherical joints allow the bogie to carry out rotational movement within required limits with respect to all planes. In addition, these can have movement in the vertical and partial displacement in the horizontal plane.

From the design point of view, the pivot assembly can consist of a pin, rigidly fixed to the main frame or car body of the locomotive on one end. On the other end, a pin is inserted in the pivot yoke, which is fixed to the frame of the bogie or the bolster.

The advantages of rigid pivot assemblies are the simplicity of their design and low cost of manufacture. Pivot assemblies, which allow lateral motions, have better dynamics in comparison with rigid joints. Furthermore, pivot assemblies with spherical joints can provide better behaviour for a locomotive in comparison with other existing designs.

3.3.3.2 Side Bearers

The main function of the side bearers is to transfer vertical loads from the car body onto the bogie frames. In addition, they should provide the ability for bogies to rotate relative to the car body and allow movements in the planes within the prescribed limits. In addition, the side bearers can generate return moments and reduce hunting oscillations of bogies, as well as provide a tilting motion of the car body, when the locomotive operates in the curved parts of the track.

Commonly used types of side bearers are:

- *Side bearer pads (rubber spring):* An example of their application is shown in Figure 3.24; inside such rubber springs, metal plates are present that separate the rubber layers; the edges of the metal plates are rubber covered in order to avoid corrosion;
- *Side bearers with return devices:* An example is depicted in Figure 3.26, showing that the design has two levels; at the top the side bearer can have a rubber or coil spring(s) and even an air spring, while at the bottom it has rollers which operate in their nest with a lubricant; the advantages of such side bearers are low coefficients of friction and the ability to get different values of return moments.

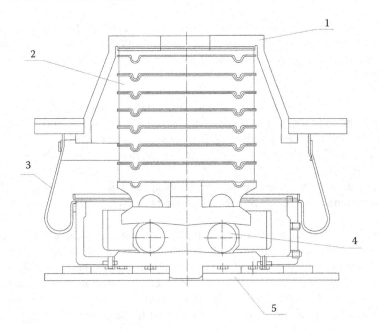

FIGURE 3.26 Side bearing with roller return device (Luganskteplovoz, Ukraine). 1—Top plate connected to an underframe; 2—rubber elements; 3—dust cover; 4—roller; 5—base plate connected to a bogie frame.

3.3.3.3 Links and Linkages

These elements can be used for different aims.

As an example, links are used in tilting trains. An example of their application is shown in Figure 3.23.

Linkages have found application in cooperation with air springs in the secondary suspension, which has transverse and lateral linkages between a bogie and a car body in order to achieve better dynamic behaviour of a high-speed rail vehicle.

3.3.3.4 Traction Rods

Traction rods are used to transfer traction and braking efforts. An example of the usage of traction rods for the connection of the pivot assembly is shown in Figure 3.24. When a powered rail vehicle is not equipped with pivot assemblies, then the traction rods can directly connect a car body and a bogie.

For damping of oscillations of traction and brake forces, traction rods can be equipped with absorbing devices; most often in such cases, rubber and rubber-metal elements or bushings have found wide application.

3.3.4 BRAKE SYSTEMS AND THEIR DEVICES

The main task of the brake system is the creation of an artificially controlled resistance force by a locomotive or train in order to control the speed or slow its movement to a full stop, and the creation of the forces which prevent the locomotive or train from inadvertent movement when it is fully stopped or parked on inclined parts of the track.

Brake systems are divided into two groups based on the method used to create the resistance force, these being either friction or dynamic.

In frictional braking systems, energy is absorbed by the friction between the wheel and the brake shoes, pads or discs, or between the rails and brake shoes in the case of rail brakes, with appropriate force loads applied on them.

Dynamical systems usually work based on the principles of transformation of kinetic energy of the train or rail traction vehicle into other types of energy (the main one being electrical) for further recovering processes and utilisation.

Based on the method of the creation of the acting control force, the brakes are divided into the following types:

- Mechanical;
- Pneumatic;
- Electric;
- Hydraulic;
- Magnetic.

The brake system of powered rail vehicles can contain several types of brakes at the same time, such as shoes or discs as well as a parking brake. In addition, it also can be equipped with dynamic, electromagnetic and rail brakes.

The main types of braking systems and their components are described in more detail below.

3.3.4.1 Basic Components of Brake Systems

The most common braking systems of rail traction vehicles which are currently in use by rolling stock operators are pneumatic systems which use shoes or wheel disc brakes. The simplified scheme of such a system is shown in Figure 3.27.

A typical brake system includes the following main components:

- Feeding and supply components (e.g. air compressor);
- Energy storage components (e.g. main and auxiliary air reservoirs);
- Acting component or actuators (e.g. pneumatic cylinders);
- Mechanical system (used for transferring of braking efforts).

In addition, a brake system includes the following elements:

- Control devices and instrumentation (driver's brake valve, emergency stop valves, etc.);
- Transportation elements (e.g. pipes, which transport air to acting devices or actuators).

Taking into account that the brake system is one of the critical systems, especially for operational safety, it also has separate pneumatic control blocks and electronic

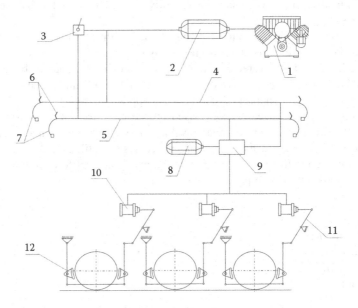

FIGURE 3.27 Typical air brake system of a rail traction vehicle. 1—Main air compressor; 2—main air reservoir; 3—automatic brake valve; 4—main reservoir pipeline; 5—brake cylinder; 6—angle cock; 7—hose; 8—auxiliary air reservoir; 10—brake cylinder; 11—lever mechanism; 12—brake block.

control systems (check of system integrity, vigilance, automatic braking and automatic control, etc.).

It is necessary to mention that a manual parking brake is also a part of the typical brake system.

3.3.4.2   Dynamic Brake Systems

The definition of a dynamic braking system covers systems that use absorption of kinetic energy of rail vehicles by means of various effects.

The main type of dynamic brake used on rail vehicles is the electromagnetic brake. This brake works on the principle of reversibility of electrical machines; it switches traction motors into generator mode. When electric currents start to be produced from such a process, the energy is absorbed from a wheel or wheels rolling on the rail, whereby the braking force is developed.

Energy obtained in such a manner can be utilised in two ways. The first way is to return it back to the power supply line; in this case, the electromagnetic brake is called a regenerative brake. Such a design solution is commonly used on electric locomotives or electric trains. Besides this, it can be used on autonomous rail vehicles which have the capability to store such energy. However, taking into account the complexity of this process, the second way has found much wider application in comparison with the first at the current time, and it is based on the principle that power obtained from a brake process is dissipated as heat in brake grid choppers or resistors. This type of electromagnetic brake is called dynamic brake or DB. In some countries, it is also known as a rheostatic brake.

On magnetic levitation transport, the current produced from the kinetic energy of the train is transmitted directly to the power line. It can be easily explained by the working principles of the linear motor.

The next type is the hydrodynamic brake; its working principles are based on the work of the friction forces arising in fluid flow. Such brakes are also often used for high-speed operations. The design of a hydrodynamic brake is most often represented as a water turbine, which is connected through the drive or mounted directly on the wheelset's axle. When braking starts, the turbine is fed with a liquid that begins to circulate on a power circuit; due to the fact that the fluid has a viscosity, there is resistance and its circulation is accompanied by heating. The generated heat is dissipated into the environment through the heat-conducting walls of the turbine.

3.3.4.3   Electromagnetic Brakes

Electromagnetic brakes use an electromagnetic force that arises by passing an electrical current through solenoids. A set of electromagnets is secured with elastic suspension on the running gear of rail vehicles (the bogie frame or the car body) and the head of the magnet is positioned over the rail head. When a current is supplied to an electromagnet, a magnetic field is produced. It moves with the rail traction vehicle and induces electric currents in the rail. As a result of the interaction of magnetic fields, eddy currents appear and they lead to the generation of an electromotive force which acts as the braking force. Electromagnetic brakes, when they are operating in normal conditions, do not have direct contact with the rail and are located at some distance from it.

3.3.4.4   Rail Brakes

A rail brake is normally a friction type of brake, wherein the braking force is generated by friction between the brake element and a rail. This type of brake is also known as a track brake. The rail brakes may be classified as the following types:

- Mechanical when the braking system is operated by mechanical lever systems;
- Pneumatic when the braking devices are actuated by a pneumatic system;
- Hydraulic when the braking devices are actuated by the hydraulic system.

Electromagnetic brakes can also be classified as rail brakes when the braking force generated by the magnetic effect is combined with the force of friction generated by the rubbing element which bears against the rail. Friction brake elements may operate on a horizontal surface of the rail head (top of rail) and on the side surface of the rail (limited application).

Unlike electromagnetic brakes which allow rigid fixation on the running gear, rail brakes are installed with elastic-damping suspension in order to reduce the shock and vibration caused by their interaction with the rail.

3.3.5   Classification of Locomotive Electric Traction

Currently, electric traction has established a dominant position in comparison with other forms of traction used in rail transport. This is due to the possibility of obtaining hyperbolic tractive effort characteristics and a wide ranging capability for the control of a traction system for various operating conditions. The latter in particular has been available since the introduction of the new generations of power semiconductor devices (transistors and thyristors) and device management based on microprocessor technology for an extended range of AC–AC and AC–DC topologies of traction. The level of development allows processing complex algorithms on locomotive control systems, including traction and braking operational modes in a real-time environment. This is related to the achievement of traction coefficients on modern rail traction vehicles of up to 50%. Currently, rolling stock uses four types of topologies for electric traction: DC, AC–DC, AC–AC with variable frequency and DC–AC. DC traction has a significant drawback of large overall dimensions of the main generator, and this is the reason why it was replaced by an alternator (a synchronous AC generator) which is significantly smaller in size. For the same reason, locomotives use AC–DC or AC–AC traction topologies. For the electric locomotives which are running on a DC electrification system, locomotives can utilise a DC–AC topology with variable voltage and variable frequency, or a DC–DC topology with pulse width control. However, it is common to call locomotives either AC or DC locomotives based on the type of electricity which is supplied to their traction motors. However, this does not indicate what other components of the traction system, such as an alternator or generator and so on, are installed in a rail traction vehicle.

This section presents a description of the most commonly used types of traction that work with DC and AC traction motors.

3.3.5.1 DC Traction

Electric traction with DC motors currently takes about 40% of the market of manufactured locomotives. The advantages of this traction system are the ease of control
of the speed and torque of DC traction motors, and the ease of switching a motor to
the generator mode for dynamic braking operations. The semiconductors required
for the production of this type of traction system can be quite simple in comparison
with AC traction systems. Controlling the rotational speed of these traction motors
can be achieved using multiple variants of their connection to the power supply
source and also different variants of the connection of their windings. DC electric
locomotives use different configurations to incorporate traction motors into the traction system:

- Series (S) connection when all the traction motors are connected in series
  provided that the voltage drop at the motor terminals is directly proportional to the number of motors used to operate at slow speeds;
- Parallel (P) connection when the traction motors are directly connected to
  the power supply source in parallel with each other—this scheme is used
  to obtain the maximum voltage at the terminals of the traction motors and,
  therefore, to achieve the maximum speed of the rail traction vehicles;
- Series to parallel (SP) connection when traction motors are connected in
  series connection to form a group—commonly, groups per locomotive are
  two or more and the groups are connected to the power supply source in
  parallel.

Speed control of DC motors is performed by the control of the armature voltage
and the adjustment of the field winding flux. DC–DC converters can be used to control both the armature and/or the field voltage. Additional control can be obtained by
using shunt resistors to adjust the field current.

The disadvantages of this type of traction system include high consumption of
non-ferrous metals, the complexity of the design of traction motors, as well as the
presence of the commutator–brush system. The commutator and brush systems limit
the maximum armature voltage and are vulnerable to flash over failures during
dynamic loading situations. Moreover, frequent service intervals with a significant
service time are required in comparison with AC induction motors. DC motors are
inferior to AC ones in terms of weight, size and price. In addition, it is almost impossible to implement a high power (>1500 kW per axle) for existing designs of DC
traction motors.

A typical scheme of electric traction with AC–DC topology of a diesel–electric
locomotive is shown in Figure 3.28.

3.3.5.2 AC Traction

The existing disadvantages of DC motors favour the development of an alternative
topology of electric traction system equipped with AC traction motors. The application of AC traction motors requires matching the level of voltage between a power
supply source and a motor, as well as the use of a sophisticated frequency converter
or inverter for adjusting the speed of rotation.

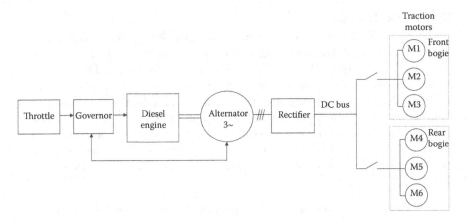

FIGURE 3.28   Example of an electric traction scheme for DC locomotive.

A typical scheme of electric traction with AC–AC topology of a diesel–electric locomotive with a traction control per bogie (bogie traction control) is shown in Figure 3.29. Another example of such a topology with a traction control per wheelset (individual wheelset traction control) is shown in Figure 3.30.

Electric traction with asynchronous traction motors is the most prevalent at the current time and it takes about 60% of the market of manufactured locomotives. This is due to the simplicity of construction, reliability and durability of this type of traction motor.

Speed and torque control of induction motors are performed by means of a pulse width modulation (PWM) inverter. For these purposes, direct torque control or vector control and its derivative strategies have found wide application. Such control approaches commonly use inverter(s) with high power semiconductors—insulated-gate bipolar transistors (IGBT) or gate turn off thyristors (GTO).

This type of electric traction allows implementing braking efforts with a large brake force down to a speed of 5 km/h. This helps reduce the cost of using other types of braking systems and also recovers more energy from the braking process if used with a regenerative braking mode.

3.4   LOCOMOTIVE DESIGN: NEW PERSPECTIVES

Increasing the role of transport in economic development and its social value requires constant improvement of vehicles, including rail traction rolling stock. The main focus should not only be on how to improve its efficiency and effectiveness, but also concentrate on safety and environmental protection.

It is possible to distinguish two currently existing directions of further research in the area of powered vehicles:

• Improvement of the traditional rolling stock;
• Investigating and developing new ways to transfer freight and passengers.

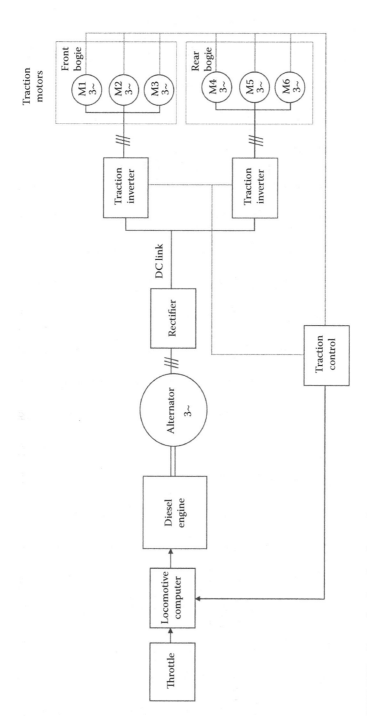

FIGURE 3.29 Example of an electric traction scheme for AC locomotive with bogie traction control.

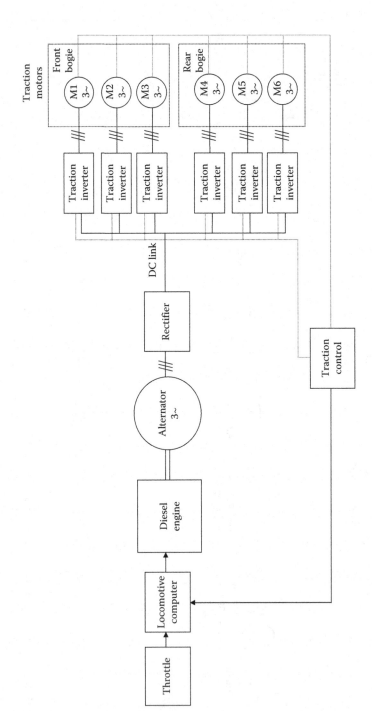

FIGURE 3.30 Example of an electric traction scheme for AC locomotive with individual wheelset traction control.

The first of these directions currently has a focus on the search for alternatives to mineral oils as an energy source for autonomous traction units, which is a dominant issue. As a result, some traction vehicles have been developed which use natural gas in a liquefied or compressed state. These locomotives have been built and are already in operation in several countries.

In addition, three of the largest U.S. rail operators, Union Pacific, BNSF and Norfolk Southern, have expressed interest and are actively cooperating with General Electric Co. and Caterpillar Inc. in working to replace all diesel traction locomotives with a more efficient locomotive model running on the more environmentally friendly liquefied natural gas fuel. Caterpillar is also planning to launch a pilot program in North America to introduce engines that use a combination of diesel and natural gas.

There is also some research and experimental design work for the application of products of coal as fuel for rail-powered vehicles. However, the large amount of emitted impurities which adversely affect the environment does not currently allow progressing these options which remain a problematic solution to the need for alternative energy sources.

The most promising option according to many experts is the application of fuel cells which convert the chemical energy from hydrogen or alcohols to produce electric power in the most direct way. Fuel cells working on hydrogen as a fuel emit the most environmentally clean combustion product in comparison with other fuels—water in the vapour state.

Atomic-powered locomotives are also a possible alternative to existing modes of rail transport propulsion. The main equipment would be very similar to that of existing locomotives, but the power source would be a nuclear reactor which heats fluid (water) to its vaporised condition before it is passed through a closed loop to and from a turbine which is connected with an electric generator. However, the introduction of nuclear power plants in rail transport is significantly constrained by security and environmental issues.

Also in the first direction, much attention is being paid to the improvement of the existing design elements and solutions in order to improve locomotive effectiveness:

- *Improved traction and braking characteristics:* To improve traction and braking characteristics, the development of new designs of running gear is being progressed which would allow a re-distribution of loads between the axles and between the wheels of one axle in real time depending on the achieved tractive efforts. In view of the fact that modern trains have almost exhausted the available adhesion coefficient between wheels and rails, there are still some works to perform to improve this. This may be achieved by using new materials for rolling surfaces by means of the introduction of new methods such as thermal, chemical and mechanical processing as well as using different physical and chemical processes with the development of new devices, which can be located on rail vehicles. Some examples of such devices already exist, but they are still not perfect solutions. These solutions may include a chemical cleaning device using various gases for rolling surfaces, which allow the removal of oxide film, thus increasing the coefficient of friction. Some devices can also clean up rolling surfaces using electric currents of high frequency. Furthermore, the creation of magnetic fields between the wheel and the rail can increase traction. However, these technologies are energy intensive to implement, and it would require extensive scientific research which must then be implemented as a separate structural solution.

It may be a good solution to use switched reluctance motors as traction motors. These motors do not have the disadvantages of those currently used, and they have low weight and exhibit load characteristics which satisfy the traction characteristics of the locomotive.

- *Reducing the impact of rail vehicles' running gear on track and improving the comfort of passengers:* Improving the design of traction gear and drives for locomotives and other motorised rail vehicles should allow further reduction of the unsprung weight of the trains and also reduce frictional losses in the gearbox, in particular through the use of motorised wheels. Improvements for operating long heavy and high-speed trains can be achieved by means of specially developed devices that monitor and then smooth the effects of longitudinal and transverse dynamics of the train.

To improve the comfort of the passengers, various methods for the reduction of excessive accelerations can be used. In addition to the standard existing solution such as dampers and tilting devices in curves, some advanced approaches for the development of fully controlled spring suspension are recommended. This requires a detailed study of how different vibrations and accelerations affect the human body. In such cases, the development of specific techniques to minimise their impact should provide better results than the averaging principle that dominates at the current time.

- *Utilisation of energy and the ability of rail vehicles to reuse it for traction:* Many research teams, as well as large industrial corporations under pressure to meet the requirements prescribed by new standards on environmental contamination and pollution, began strongly developing technologies for the creation and production of devices which can re-utilise energy derived from braking modes of rail vehicles. The fuel obtained from such a technological process can be re-used again in diesel engines or fuel cells. However, such technologies have a low efficiency in comparison with hybrid technologies which use super capacitors, batteries or fly wheels.

When operating a conventional rail vehicle at speeds up to 200 km/h, it is possible to see that about 40% of the energy produced by the propulsion plant is consumed for traction (the kinetic energy of vehicles), another 40% of the energy is consumed by natural oscillations of the running gear and their interconnections, a further 10% is spent to overcome the frictional forces in the systems and components of vehicles and the interaction forces between wheels and rails, while the final 10% is needed to overcome air resistance. Therefore, it would be a significant advantage to be able to use the conversion of the natural oscillations of vehicle parts as an additional source of energy. Such an approach will reduce the heating of the environment in comparison with heating generated by modern damping systems.

The application of electric generators as elements of such new systems should also allow achieving precise damping characteristics. The downside of this trend is the increased consumption of electrical steels and non-ferrous metals, as well as the usage of complex control systems for suspension and coupling of rail vehicles, which would substantially increase their cost, but would also significantly increase the efficiency of the utilisation of the energy produced by the propulsion plant.

At the current time, the second direction is based on the design of fundamentally new rolling stock at what is considered to be the conjunction of rail, road, air and pipeline transport modes:

- *Buses with rail wheels and vehicles for combined usage:* The development of such vehicles has arisen due to the possibility of using existing infrastructure, such as roads and railways. These types of traction vehicles may promote the selection of the shortest routes. For example, in the absence of a suitable road surface, it is possible to use a rail track in order to reach the destination point for passengers and goods. That is a very common difficulty for areas with a high density of population. This technology is convenient for passengers since the movement of such vehicles is direct and without additional transfers.
- *Levitation train on cushion of air:* In such trains, an aerodynamic suspension is used instead of a magnetic one. This technology is also called ground effect transport system (GETS). To provide the propulsion in this case, linear motors and turbines with electric transmission can be used. The main disadvantage of GETS is that it can be affected by extreme weather conditions.
- *Vacuum train:* These trains, also known as 'vactrains', will allegedly reach speeds of 6400–8000 km/h—that is 5–6 times the speed of sound in air, and these trains would move very quickly due to the lack of aerodynamic drag. For trains of this type, it is necessary to build a tunnel with a vacuum system and vacuum support stations, which at the moment would be technically difficult and a quite expensive matter. In this regard, China, Japan, Switzerland and the United States are implementing projects that would not use the full vacuum but only partially rarefied air in tunnels or pipes. In that case, such trains could reach speeds of 1000–1500 km/h.

Summarising the above, it is necessary to say that the cost of creation and development of new types of rail vehicles and trains, as well as their infrastructure, is extremely high. However, advantages in speed, reliability and cost of transport show that these directions will progress and some may eventually replace the existing types of transport.

4 General Modelling Techniques

Rail vehicle modelling encompasses a wide range of analysis that can be performed. This includes modelling of the dynamic motion, pneumatic brake system, thermal modelling of the brakes, aerodynamic analysis, fatigue and failure analysis. The rail vehicle is also affected by outside influences such as the track stiffness and the rail profile. It may be necessary to model these additional systems to accurately model the rail vehicle behaviour. Where possible it is better to simplify models to reduce the modelling effort required and the computational requirements. It should be recognised that rail vehicles are used in trains made up of many vehicles, and complex vehicle models may not be feasible when modelling complete trains.

The railway environment is full of other modelling possibilities that are outside the context of this text but still affect the rail vehicles. These include the modelling of the power grid which dictates the power available to electric locomotives and regenerative possibilities, weather modelling to study track buckling and wind loadings, rail network traffic modelling and scheduling, passenger and cargo/material flows and even train crewing scheduling and the timing of maintenance activities.

4.1 DYNAMICS OF BODIES

The dynamic modelling of railway vehicles and trains use the same techniques used for the modelling of most mechanical systems as detailed in [1,2]. This topic is covered in depth in the next chapter. Dynamic bodies that make up rail vehicles are modelled based on their mass and inertial characteristics. Connections between wagon bodies, bogies, wheelsets and rails are modelled as springs, dampers and non-linear force elements. Due to the high number of non-linearities present in rail vehicles, numerical solvers are typically used. Linear analysis where the motion is solved by traditional methods using differential equations can provide insight into the natural frequencies and the frequency response behaviour of the various dynamic vibration modes of a rail vehicle. While this information is valuable, the main interest to rail engineers is the expected time series displacement and force responses that will be experienced by the rail vehicle during normal operation.

4.1.1 RAIL VEHICLE

The rail vehicle is made up of different components which can be modelled in the system as individual bodies. Figure 4.1 shows the bodies of a freight wagon model, and of the supporting track and sub-ballast. The connections are represented as vertical springs but these connections could also include non-linear connections and

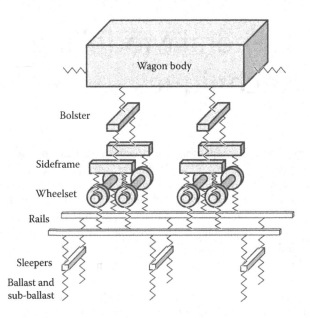

FIGURE 4.1   Freight wagon model.

connections in the other linear and rotational directions [3–6]. The connections between the wagon body and other rail vehicles are shown, but there could be other external forces on the wagon body such as wind loading (and pantograph forces if the vehicle was an electric locomotive). Detailed descriptions of the connections are provided in Chapter 5.

Some major simplifications can be made to the model if modelling only the bounce, pitch and roll motions. This is possible as the majority of the freight wagon connections are steel on steel connections. The simplifications can be used to determine the fundamental vibration modes that can exist in the wagon [7]. As the damping in the bolster to sideframe connections in the three-piece bogie consists of friction wedges, the resulting damping can be considered as Coulomb damping and, for this reason, the natural vibration frequency will be the same as the undamped natural frequency. For all fundamental modes, the bolster mass and inertia can be combined with the wagon body. The sideframe mass and wheelset mass can also be combined together if there is no primary suspension. Additionally, if the rail is considered as a stiff element, then the sideframe and wheelset masses are supported solidly by the track structure and would not contribute to any vibration of the vehicle.

The bounce mode can be modelled as a single-degree-of-freedom system as shown in Figure 4.2. The spring represents the equivalent spring from combining all the sideframe to bolster spring nests. If it was desired to determine the amplitudes of vibrations, the constant Coulomb friction damping would need to be added to the connection and a vertical displacement added to the sideframe and wheelset body.

The simplified rail vehicle pitch model groups the front and rear sideframe to bolster spring nests, and considers only the rotational motion of the rail vehicle body as shown in Figure 4.3.

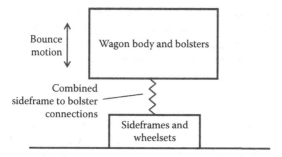

FIGURE 4.2 Simplified rail vehicle bounce model.

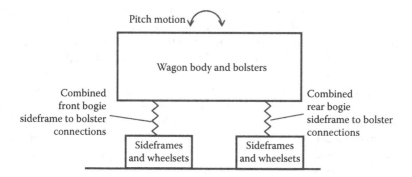

FIGURE 4.3 Simplified rail vehicle pitch model.

The roll vibration mode can similarly be modelled by combining the left and right spring nests into their respective equivalent springs as shown in Figure 4.4. The simplified roll model may be an oversimplification as it assumes the sidebearer connections between the bolster and the rail vehicle body are solid connections. In reality, these connections can contain spring stiffness and damping; also, if there is clearance in the connections, the rail vehicle body may have motion similar to an inverted pendulum.

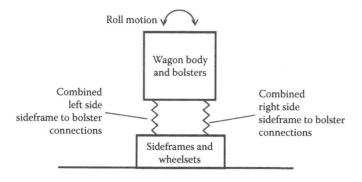

FIGURE 4.4 Simplified rail vehicle roll model (end view).

While these simplified models give some insight, analysis of a full rail vehicle model is needed to provide the details of rail vehicle performance for different track perturbations and different coupler and external force events. Full models require non-linear solvers which are described further in this chapter.

4.1.2 TRAIN

Train modelling involves connecting rail vehicle and locomotive models together to create a multiple-degree-of-freedom model as detailed in [1]. Owing to the high number of degrees of freedom and non-linearities in the vehicle connections, numerical methods such as Runge–Kutta are used to simulate the response. Generally, simplified vehicle models are used to reduce the computational power required to solve the models [8,9]. The most common simplification used is where only the longitudinal stiffnesses and forces are considered and each rail vehicle is modelled as a single mass. This means ignoring the vertical stiffnesses and connections in the rail vehicle models as shown in Figure 4.5. Train modelling is covered in-depth in Chapter 6.

The output from such models provides longitudinal in-train forces, rail vehicle body accelerations, coupling displacements and train speed. The accuracy of the transient longitudinal forces and accelerations is highly dependent on the accuracy of the coupler model. Train models are generally analysed in the time domain as this is useful to assess the track location at which maximum longitudinal forces occur as well as the suitability of the locomotive power for a particular train length. Fatigue analysis is another important aspect of longitudinal train modelling and is focused on determining the fatigue life of coupling components and, to a lesser extent, the rail vehicle frame and/or body [10,11].

There are a number of longitudinal force inputs on the rail vehicles that need to be considered in the train model, including the locomotive traction and braking forces, wagon braking forces, track grade forces, curving resistance, rolling resistance and drag due to aerodynamic performance of the train, and the current wind speed and direction. Locomotive tractive and braking forces are dependent on the driving control inputs and the locomotive type. Rolling resistance forces include the drag due to the bearings as well as the wheel–rail drag force. Rolling resistance can vary significantly through curves, particularly when large longitudinal forces are placed on rail vehicles causing high lateral wheel loads [12]. While longitudinal train models are focused on determining the longitudinal forces on the rail vehicles, the lateral rail

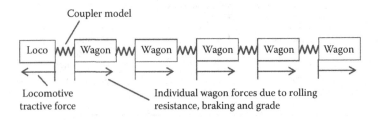

FIGURE 4.5 Longitudinal train model.

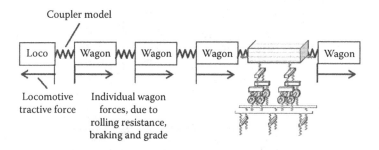

FIGURE 4.6 Detailed rail vehicle model in a longitudinal train model.

vehicle force can be determined from the speed of the rail vehicle, the radius of the curve and the longitudinal force on the rail vehicle [13].

Accuracy of the train speed from a train model is dependent on many variables such as the locomotive engine model, brake models and rolling resistances. For this reason it is difficult to determine an accurate indication of train speed that would occur in real train operation. Comparing a train model's speed profile with an actual train's speed profile will invariably show differences that will be more apparent towards the end of the journey due to cumulative error effects. When using actual locomotive control actions in a train simulation, methods have to be adopted to ensure the track position at which control changes occur are the same for both the actual operation and the simulation. Actual field test data should be used to verify and refine train models [14,15].

To better understand the behaviour of a rail vehicle when it is travelling in a train, some different techniques can be used. For long trains it is impractical to create a train made up of detailed rail vehicle models, but detailed rail vehicle models can be added to the train model at the points of interest as shown in Figure 4.6. In some cases, multiple detailed rail vehicle models may need to be added to the train model to reduce end effects, as the simple longitudinal train model may not accurately calculate the lateral coupler forces that are transferred to the rail vehicle under study. An alternative way to more accurately model the behaviour of a single wagon in a train is to run a longitudinal train model, then use its longitudinal force results for the rail vehicle position of interest and apply these to a detailed rail vehicle model [16].

4.2 WHEEL–RAIL CONTACT PATCH

The wheel–rail connection is a very important part of modelling rail vehicles. The contact patch typically forms an elliptical area where the wheel touches the rail and transfers longitudinal, vertical and lateral forces. The curvature of the wheel and rail creates high stresses within the contact patch, causing plastic deformation and thus work hardening of the rail and wheel, and this can result in surface and sub-surface fatigue cracks as shown in Figure 4.7.

In normal centre tracking conditions, the wheel tread and rail contact at a single contact patch. When there are high lateral forces, the wheelset can be forced so that the wheel flange also contacts the rail, resulting in a two-point contact [17]. The location

FIGURE 4.7   Railhead surface fatigue cracks.

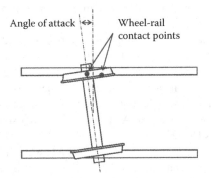

FIGURE 4.8   Two-point contact and angle of attack.

of the flange contact is also dependent on the angle of attack of the wheelset as shown in Figure 4.8. Two or more points of contact can also occur depending on the wheel–rail profile design and the degrees to which the wheel and rail profiles are worn. Other cases of multiple contact points occur when traversing through points and crossings as the wheel crosses over from one rail to another [18–20]. The contact patch location is determined from the relative position of the wheelset in relation to the railhead and the condition of the wheel and rail profiles. In wheel–rail models, the contact force is typically determined from Kalker's rolling contact model [21], the Heuristic non-linear creep force model or Polach's non-linear model [22,23].

When the wheel and rail profiles are very similar, conformal contact can occur and many points of contact result. Some dynamic railway vehicle modelling software can model multiple-point contact as shown in Figure 4.9 [24,25].

4.3   BRAKE MODELLING

The two main types of brakes used in rail vehicles were introduced in Chapter 2. The most difficult brake type to model is the traditional fully pneumatic air braking as

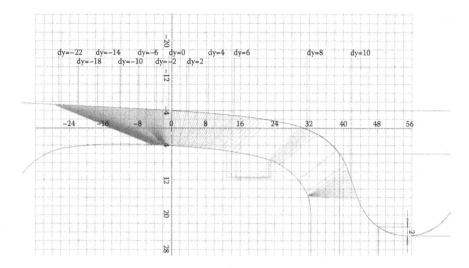

FIGURE 4.9 Multiple-point wheel–rail contact model.

the timing of the brake application on each vehicle is dependent on the pressure drop in the long brake pipe that extends down the length of the train [26,27].

4.3.1 PNEUMATIC BRAKES

Pneumatic brakes or 'air brakes' operate by use of a long 'brake pipe' that runs down the length of the train. The brake pipe supplies air to a reservoir on each rail vehicle (except in some designs where the braking systems on permanently coupled wagons share reservoirs), and the pressure in the pipe acts as a signal to apply and release the brakes. The brake model is made up of two integrated pneumatic systems, one for the brake pipe and the other for the brake valve and reservoir on each rail vehicle as shown in Figure 4.10. Fluid modelling techniques can be used to model the operation of these two combined systems [28–31]. Alternatively, empirical lookup tables are sometimes used to reduce the computational power required for train simulations [32]. An empirical lookup table is typically created from experimental brake tests and is only valid for that particular train configuration, although models are sometimes modified for use with different train lengths. Empirical models are also only valid for the brake application amplitudes that have been measured. With railway brakes there are three main levels of brake application, these being minimum, full service and emergency. Various other 'service' brake levels can range in magnitude from the 'minimum' application to the 'full service' application. An appropriate

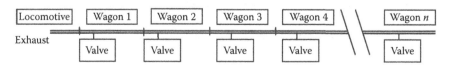

FIGURE 4.10 Train pneumatic brake model.

number of brake tests should be done to cover a suitable number of applications. Empirical models may not accurately model the brakes when changes are made to the level of braking throughout the application. This may occur when an initial brake application is made and then the brake application level is subsequently increased or reduced. Empirical models are best used for modelling where only set brake applications are made with no changes in the braking level during the brake application.

Pneumatic rail vehicle brake systems can be quite intricate, with many reservoirs and valves. An example of a brake system that is used on heavy haul wagons is shown in Figure 4.11. Inside the triple valve there are more valves, chambers and chokes. While the air brake valve arrangement started its development as a simple system, more features have been continually added throughout its design lifetime to improve its performance. The general operation of the system is where the brake pipe is normally at a set pressure which charges the rail vehicle reservoirs, and a reduction in brake pipe pressure triggers the valve to actuate the brake cylinders. As the brake pipe is returned back to its maximum pressure, the rail vehicle brakes release.

The brake valve can be modelled as a fluid system with the various valves and reservoirs. However, for the study of train dynamics, a mixture of fluid modelling and empirical modelling may be adequate to provide an accurate value for the final braking force. The internal operation of the brake valve is of little or no interest in the study of train dynamics. If an empirical model for the brake system is used, it is possible to include the entire system in the empirical model and thus eliminate the need to model each rail vehicle brake system individually. Fluid models of the brake pipe and rail vehicle brake systems are useful when studying how the brake system performance is affected by any changes to the brake system. What is not shown in the rail vehicle brake schematic is the load switch which causes a smaller braking force to be applied to empty or lighter rail vehicles. This is done to eliminate the possibility of wheel slip during braking. As the load switch is a simple on/off switch or a variable switch, this is relatively simple to model using the logic functions.

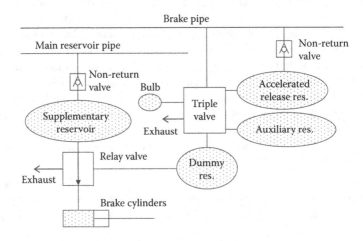

FIGURE 4.11 Rail vehicle pneumatic brake model.

4.3.2 Electronically Controlled Pneumatic Brakes

Electronically controlled pneumatic brakes are commonly referred to as ECP brakes. ECP brakes use an electronic signal to apply and release the brakes. Similar to a traditional air brake system, with an ECP system the brake pipe is still used to charge the rail vehicle air reservoirs. A set of electronic valves on each rail vehicle control the filling of the brake cylinders to apply and release the brakes. ECP allows the brake application to be applied to each rail vehicle simultaneously, although some delays in filling and releasing the brake cylinders may be added to the electronic valves to reduce the longitudinal force transients. These aspects are modelled in time-based simulations, using either logic functions or simple fluid models of the pressurised air transfer from the reservoirs to the brake cylinders [33,34]. A pneumatic brake pipe should be modelled in cases where the train is very long or there are frequent brake applications and releases. Both of these situations may cause insufficient brake pipe air flow to be available to keep reservoirs adequately charged. In such systems, the inclusion of a brake pipe model might be needed to provide an accurate ECP brake system model.

4.3.3 Wheel Brakes

Wheel brakes use the brake shoes that are applied to the outside of the wheel treads as shown in Figure 4.12. Heat caused by the brake applications is dispersed into the wheelset. Lengthy continuous brake applications, such as those made on long steep downgrades, have the potential to increase the temperature of the wheel tread to a temperature that causes brake fade. During brake fade, the coefficient of friction reduces along with the applied braking force. Suitable thermal models may need to be created to accurately determine the effect of brake fade [35–37]. As the brakes are on the outside of the wheel running surface, there is also the possibility of changes to the coefficient of friction due to moisture and other contaminants [38–40]. Other contaminants may include rail lubricant that is primarily used to reduce wheel wear and/or lateral wheel forces. These aspects of the brake model should be considered with regard to the operating environment.

FIGURE 4.12   Wheel brakes.

4.3.4 DISC BRAKES

Disc brakes are typically used on modern passenger wagons to ensure consistent braking performance. Disc brakes can be affected by brake fade as previously mentioned for wheel-type brakes [41]. With disc brakes, the braking force is typically applied using pneumatic cylinders similar to the application of wheel brakes. Disc brakes are fitted to the axle or to the sides of the wheels, and are therefore more protected from contaminants such as moisture and rail lubricant. Modelling of the disc brakes can follow a similar procedure to that used for wheel brakes, using thermal models to predict changes in friction coefficients as in [42].

4.4 AERODYNAMICS

Drag caused by wind resistance can be included into the general rolling resistance formula. This formula includes the speed of the rail vehicle. Wind resistance will be dependent on the distance the rail vehicle is from the front of the train and the shape of the rail vehicle and preceding rail vehicles [43]. In unit train operations, the wind resistance variation is considered to be small and all rail vehicles are given the same wind resistance values. The same could be said for mixed freight trains unless there is significant variation in the rail vehicle types. The consideration of aerodynamics is also important in calculating the wind overturning probabilities [44,45]. For simple flat-sided rail vehicles, general wind pressure loadings could be calculated. For more complicated rail vehicle shapes, fluid modelling software could be used [46,47]. Aerodynamic modelling becomes more crucial as the speed of the train increases. This includes cases where the train enters tunnels and where high-speed trains pass each other on adjacent lines [48–50]. Platform safety is another important aspect if trains are allowed to pass through passenger platforms at high speeds. If a person is standing too close to the platform edge, the air vortexes of the train have the possibility of causing the person to fall onto the track. Also, as a high-speed train enters a tunnel, a pressure shockwave is produced that can damage passenger ear drums.

4.5 INTRODUCTION TO FEM

The finite element method (FEM) is a numerical technique that is used to reduce a modelling problem down into small elements. These small elements are modelled as simple modules with external inputs from adjacent elements. A comprehensive description of FEM is given in [51]. FEM can be used to solve complex systems with good accuracy. FEM models consist of finite elements connected to other finite elements at points called nodes. An example of a 2D structural FEM model is shown in Figure 4.13. Types of FEM that can be used in railway applications include modelling the rail vehicle structure, rail, air brake system, heat transfer due to braking and movement of fluids in tank wagons [52].

FEM includes two types of modelling, time independent and time dependent. Time-independent models treat the problem as a static problem and do not consider the inertia of the elements. These models can be used to determine the stresses and strains that occur in rail vehicle components due to externally applied forces. Time-dependent

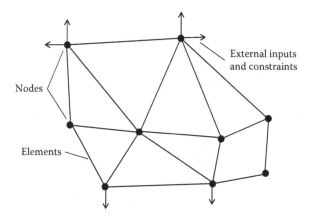

FIGURE 4.13 2D FEM elements and nodes.

models are used to model the forces and stresses due to the dynamics of the elements. Two methods are used in time-dependent FEM, explicit integration and implicit integration as explained by [53]. Explicit solvers calculate the state of the model for the next time step using only the present state of the model. The stability of explicit solvers is dependent on the step size which needs to be based on the change of state speed or wave through the model. An example of an explicit solver is that which is used in the ABAQUS software package when using the ABAQUS/EXPLICIT solver. Implicit solvers use both the present and approximated future states. The application of implicit solvers is computationally more expensive due to the non-linear nature of the step calculations, but it is possible to use larger time steps (noting that implicit solvers often use predictor–corrector methods, using more and smaller steps than the initial step size). Implicit solvers are inherently stable due to the iteration process. A commonly used implicit method is the Newmark method, which uses present and future approximations of displacement, velocity and accelerations in the iterative process [54].

4.6 FEM OF RAIL VEHICLE STRUCTURE

Modelling of the rail vehicle structure can be handled in two ways. For typical dynamic modelling, the vehicle structure can be considered either as a rigid body or as a single stiffness [1,2]. This allows simplified longitudinal train and vehicle dynamics modelling and determination of the forces on the rail vehicle. The external forces can then be applied to a detailed FEM model of the rail vehicle body/frame to determine the expected stresses and fatigue lives [55,56]. The FEM mesh of the structure is sized according to the accuracy desired, the loading type and the computational speed required. Figure 4.14 shows an FEM model of part of a rail vehicle frame modelling an anti-ride-up device. There are numerous structural FEM software packages on the market that can be used in the analysis, and the choice is based largely on ease of use, support available, accuracy and the pricing structure. FEM packages may also provide a way to determine the simplified vehicle stiffnesses that can be used in the dynamic analysis by either a dynamic train model or a detailed rail vehicle dynamic model.

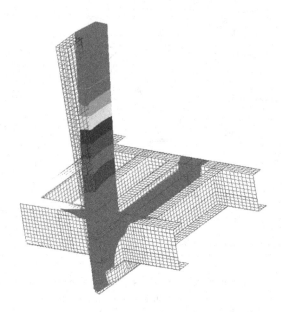

FIGURE 4.14   FEM of rail vehicle frame and anti-ride-up device.

The second method of modelling the rail vehicle structure is to use software that incorporates both an FEM of the vehicle as well as the dynamic model [57]. This approach is very similar to the first approach, but it is simplified as the inputs and outputs of the FEM and dynamic models are handled by the software. This type of modelling is particularly suited to crash simulations where the deflection of the rail vehicle body and frame affect the vehicle dynamics [58–61]. To reduce the computational effort required, the FEM models integrated with dynamic modelling software are generally simplified. As a result, detailed FEM rail vehicle body analysis typically uses standalone packages to increase the accuracy of the results. Models integrating vehicle dynamics and FEM can still be essential to get more accurate estimated forces where forces are modified by flexure or deformation in solid body components.

4.7   FEM OF RAIL

Finite element modelling of rail is performed to determine the stresses and/or the deflections in the rail. Similar to FEM of the vehicle body/frame, the FEM of rail can be done separately or combined with dynamic models [62]. Dynamic FEM rail models are useful to determine the stresses in the rail when traversing over joints and other non-continuous sections of rail [63]. Insulated rail joints which are installed to provide track circuits for train detection with signalling systems are found in most modern railway networks. Additional stresses occur at these joints, causing wear and early failure due to the discontinuity of the rail structure. An insulated rail joint is shown in Figure 4.15.

For continuous rail sections that do not have any defects or discontinuities, FEM of the rail can be performed independently of the dynamic solvers [64]. The wheel and rail needs to be meshed appropriately to accurately determine the stresses for the

FIGURE 4.15 Insulated rail joint.

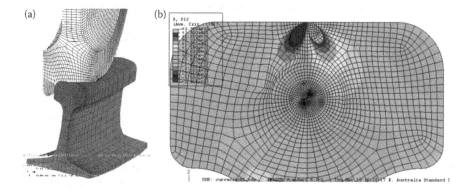

FIGURE 4.16 FEM meshed model examples. (a) Wheel and rail, (b) railhead.

type of loading that is encountered. Examples of meshed wheel and rail FEM models are shown in Figure 4.16.

FEM of the rail track where the rail can be considered as a flexible beam supported between the sleepers is another application of rail FEM. This area is included in the next section.

4.8 RAIL TRACK, SUB-STRUCTURE AND BRIDGE MODELLING

In most situations, a well-maintained track will have a minor effect on the dynamics of the rail vehicle, and the rail can be effectively modelled as a rigid element as explained in [65]. However, the track structure may have an effect when its stiffness is low due to deficiencies in the ballast or sub-ballast and when there are changes in the track stiffness. High wheel loading coupled with the resonance frequencies of the rail vehicle can also require modelling of the rail stiffness. As presented in the example of the full freight wagon model of Figure 4.1, the track structure has an effect on the dynamics of rail vehicles and can be modelled

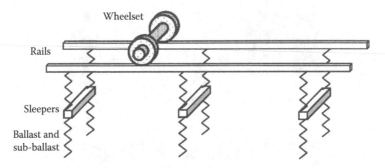

FIGURE 4.17   Track model components.

[66,67]. The three main elements of the model include the rail stiffness, the connection between the rails and sleepers, and the ballast and sub-ballast characteristics as shown in Figure 4.17.

Rail stiffness between the sleepers can be modelled as simple beam elements. Modelling of the rail stiffness is important when the speed of the rail vehicle and the spacing of the sleepers coincide with the resonance frequencies of the rail vehicle. To investigate these interactions, the track model and vehicle model need to be combined into one model. Stiffness between the sleepers and rail is dependent on the pads placed at these locations [68]. Generally the stiffness of these elements is high enough that it does not warrant any consideration, but in some instances such as noise mitigation, softer elements are used which have a greater effect on the vehicle dynamics. Ballast effects on vehicle dynamics are most noticeable when ballast defects occur, such as at 'mud holes' where the track 'pumps' due to the lower than normal stiffness [69]. These occurrences can cause rail vehicle bounce and pitch, and these effects are magnified if the rail vehicle transverses the location at a speed that coincides with the resonance bounce or pitch frequency of the rail vehicle. An example of a track location at which track 'pumping' occurs is shown in Figure 4.18. The 'pumping' or large vertical movement of the track is distinguished by the lighter colouring in the ballast where the ballast stones are moving due to the track motion and creating dust through abrasion with other stones and the surface of the concrete sleepers [70].

FIGURE 4.18   Track pumping.

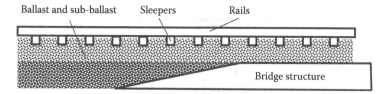

Ballast and sub-ballast   Sleepers   Rails

Bridge structure

FIGURE 4.19 Gradual lead-in to bridges and culverts.

Figure 4.18 also shows another important aspect of track modelling, and that is where there is a large change in track stiffness caused by the track transitioning from a ballasted track sub-structure to a more rigid bridge or culvert sub-structure (the latter can be seen in Figure 4.18). In an attempt to reduce the dynamic effects on rail vehicles due to these changes, a gradual transition is sometimes provided by using a foundation that extends out prior to the bridge or culvert as shown in Figure 4.19. The dynamic model of the bridge structure can be performed by structural FEM analysis packages, although these are typically modelled just as a change in the sub-grade stiffness [71–74].

4.9 PANTOGRAPH MODELLING

The overhead electrical catenary can be modelled as flexible elements fixed at set intervals that coincide with the spacing of the support structures [75]. The catenary tension, which relates to the effective vertical stiffness of the catenary, is provided by weights hung at regular intervals. Pantographs connected to rail vehicles can be modelled as a mass–spring–damper system [76]. The overhead catenary, pantograph and vertical rail vehicle motion needs to be modelled as a complete system to determine the interactive forces between the catenary and the pantograph. These forces are important so that a constant contact is maintained and the force is not excessively large to ensure the minimum wear rates of the catenary and pantograph. Wind loading is an important factor as this affects the positioning of the catenary [77,78]. The pantograph–catenary is a three-dimensional system as the contact wire is positioned so that it moves laterally across the pantograph to ensure constant wear of the pantograph. However, the dynamics of the rail vehicle are typically not affected by the catenary due to the small forces involved when compared to the large mass of the rail vehicle. Modelling of the pantograph is primarily performed to ensure that the spring rate and damping used in the pantograph are sufficient to keep contact with the contact wire in all cases of expected track perturbations. In this respect, only a vertical model is required and the system can be reduced to a one-degree-of-freedom model as shown in Figure 4.20. Changes in the vertical movement of the rail vehicle body and vertical movement of the contact wire will affect the force between the pantograph and the contact wire [79].

4.10 MODELLING TECHNIQUES

As mentioned throughout this chapter, there are numerous modelling techniques that can be applied to rail vehicles. It may even be possible for the performance of rail

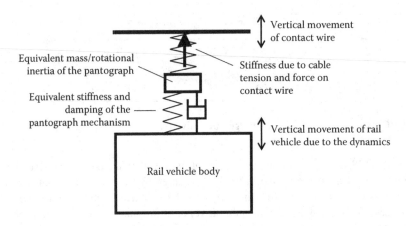

FIGURE 4.20   One-degree-of-freedom pantograph model.

vehicles to be determined using quasi-static analysis [80]. For dynamic motion, linear dynamic models can be applied and these will provide some information about the behaviour of the rail vehicles. More complex multi-dimensional dynamic non-linear models are needed to fully model the complete operational range. Modelling techniques range from the simple solution of differential equations to the time-based numerical integration solutions. Software packages provide various modelling tools such as FEM software for the analysis of stress and fatigue and dynamic modelling software [81,82]. Some software suites provide a seamless way to incorporate both FEM and dynamic models. The choice of software is important depending on the investigation required. General dynamic modelling software will be more flexible, but it will take more time to develop the necessary models [83]. Rail-specific modelling software has advantages as they generally provide suitable connection elements. How flexible these models are depends on the software. In some cases, it may only be possible to model certain aspects by either using multiple software packages or via the commissioning of tailor-made software. General modelling techniques are well described, and it is possible to use software programming languages to perform basic modelling. More specific discussion of railway modelling software is provided in the following chapters.

4.10.1 DYNAMIC MODEL ELEMENTS

The basic dynamic model elements consist of masses that are connected together. The masses have linear and rotational inertias that resist acceleration. The motion of these bodies is dictated by the net applied force on each body. The forces applied to the dynamic bodies can be due to external forces, movements and the movement relative to other connected bodies. Some of the dynamic model elements that relate to rail vehicles have been presented in the previous sections, and these include masses, springs, viscous and friction dampers. More specialised connection elements include the wheel–rail connection elements. Differential equations of motion can be formulated for each possible degree of freedom for each dynamic body.

4.10.2 NUMERICAL INTEGRATORS

To solve non-linear differential equations, numerical integrators can be used such as the Taylor series or Runge–Kutta methods [84,85]. However, as this method of solution is not exact, the accuracy of the results needs to be evaluated. Accuracy is highly dependent on the time step used and it is possible that, if large time steps are used, the solution may not even converge. The disadvantage of using small time steps is that computational time will be increased. Most rail vehicle software packages generally automatically adjust the time step and also have a provision to manually set the time step. To ensure the validity of the simulation solvers, it may be necessary to reduce the time step and check that the results do not change significantly. If the results do change, the time step should be reduced further to find where the model is stable. Alternatively, small changes to non-linear elements (slight changes to stiffnesses, more transition points at changes in slopes etc.) can greatly improve stability without changing model integrity.

REFERENCES

1. S. Iwnicki, *Handbook of Railway Vehicle Dynamics*, CRC Press, Taylor & Francis Group, Boca Raton, FL, 2006.
2. V.K. Garg, R.V. Dukkipati, *Dynamics of Railway Vehicle Systems*, Academic Press, New York, NY, 1984.
3. B.M. Eickhoff, J.R. Evans, A.J. Minnis, A review of modelling methods for railway vehicle suspension components, *Vehicle System Dynamics*, 24(6–7), 1995, 469–496.
4. R.F. Harder, Dynamic modeling and simulation of three-piece freight vehicle suspensions with nonlinear frictional behaviour using Adams/Rail, *Proceedings of the 2001 IEEE/ASME Joint Railroad Conference*, Toronto, Canada, 17–19 April 2001, pp. 185–191.
5. J.F. Gardner, J.P. Cusumano, Dynamic models of friction wedge dampers, *Proceedings of the 1997 IEEE/ASME Joint Railroad Conference*, Boston, MA, 18–20 March, 1997, pp. 65–69.
6. S. Bruni, J. Vinolas, M. Berg, O. Polach, S. Stichel, Modelling of suspension components in a rail vehicle dynamics context, *Vehicle System Dynamics*, 49(7), 2011, 1021–1072.
7. J. Evans, B. Mats, Challenges in simulation of rail vehicle dynamics, *Vehicle System Dynamics*, 47(8), 2009, 1023–1048.
8. M. Ansari, E. Esmailzadeh, D. Younesian, Longitudinal dynamics of freight trains, *International Journal of Heavy Vehicle Systems*, 16(1–2), 2009, 102–131.
9. Z. He, S. Luo, W. Ma, Research on the influence of coupler on heavy haul locomotive dynamics, *Proceedings of the International Conference on Computer Design and Applications*, Qinhuangdao, China, 25–27 June, 2010, Vol. 3, pp. 283–290.
10. J. Fang, W.Z. Zhao, J. Zhang, Fatigue life prediction of CRH_3 carbody based on rigid-flexible coupling model and the master S-N curve, *Proceedings of the 1st International Workshop on High-Speed and Intercity Railways*, Shenzhen and Hong Kong, China, 19–22 July 2011, pp. 403–412.
11. G. Li, M. Ha, Study on the load spectrum of the coupler device of the heavy haul freight electric locomotive running at 120 km/h, *China Railway Science*, 32(1), 2011, 86–90.
12. K. Kim, S. Chien, Simulation-based analysis of train controls under various track alignments, *Journal of Transportation Engineering*, 136(11), 2010, 937–948.
13. C. Cole, M. McClanachan, M. Spiryagin, Y.Q. Sun, Wagon instability in long trains, *Vehicle System Dynamics*, 50(Suppl), 2012, 303–317.

14. E. Kassa, J.C.O. Nielsen, Dynamic interaction between train and railway turnout: Full-scale field test and validation of simulation models, *Vehicle System Dynamics*, 46(Supp1), 2008, 521–534.

15. V.N. Rangelov, Gradient modelling with calibrated train performance models, in *Computers in Railways XIII: Computer System Design and Operation in the Railway and Other Transit Systems*, C.A. Brebbia, N. Tomii, J.M. Mera, B. Ning, P. Tzieropoulos (Eds.), WIT Press, Ashurst, UK, 2013, pp. 123–134.

16. Z.Q. Xu, S.H. Luo, Q. Wu, W.H. Ma, The curving behaviour of heavy-haul locomotive couplers, *Applied Mechanics and Materials*, 197, 2012, 409–414.

17. J. Piotrowski, H. Chollet, Wheel–rail contact models for vehicle system dynamics including multi-point contact, *Vehicle System Dynamics*, 43(6–7), 2005, 455–483.

18. E. Kassa, C. Andersson, J.C.O. Nielsen, Simulation of dynamic interaction between train and railway turnout. *Vehicle System Dynamics*, 44(3), 2006, 247–258.

19. C. Andersson, T. Dahlberg, Wheel/rail impacts at a railway turnout crossing, *Rail and Rapid Transit*, 212(2), 1998, 123–134.

20. M.J.M.M. Steenbergen, Modelling of wheels and rail discontinuities in dynamic wheel-rail contact analysis, *Vehicle System Dynamics*, 44(10), 2006, 763–787.

21. J.J. Kalker, Wheel–rail rolling contact theory, *Wear*, 144(1), 1991, 243–261.

22. J. Pombo, J. Ambrósio, M. Silva, A new wheel–rail contact model for railway dynamics, *Vehicle System Dynamics*, 45(2), 2007, 165–189.

23. N. Bosso, M. Spiryagin, A. Gugliotta, A. Somà, Review of wheel–rail contact models, in *Mechatronic Modeling of Real-Time Wheel–Rail Contact*, Springer, Berlin, 2013, pp. 5–19.

24. I. Coleman, E. Kassa, R. Smith, A multi-point contact detection algorithm combined with approximate contact stress theories, *Proceedings of the 1st International Conference on Railway Technology: Research, Development and Maintenance*, Tenerife, Canary Islands, Spain, 18–20 April 2012, pp. 1986–2001.

25. J.C.O. Nielsen, A. Ekberg, Acceptance criterion for rail roughness level spectrum based on assessment of rolling contact fatigue and rolling noise, *Wear*, 271(1–2), 2011, 319–327.

26. A. Nasr, S. Mohammadi, The effects of train brake delay time on in-train forces, *Rail and Rapid Transit*, 224(6), 2010, 523–534.

27. M.A. Murtaza, S.B.L. Garg, Brake modelling in train simulation studies, *Rail and Rapid Transit*, 203(2), 1989, 87–95.

28. L. Pugi, M. Malvezzi, B. Allotta, L. Banchi, P. Presciani, A parametric library for the simulation of a Union Internationale des Chemins de Fer (UIC) pneumatic braking system, *Rail and Rapid Transit*, 218(2), 2004, 117–132.

29. L. Pugi, D. Fioravanti, A. Rindi, Modelling the longitudinal dynamics of long freight trains during the braking phase, *Proceedings of the 12th IFToMM World Congress in Mechanism and Marine Science (CD)*, Besançon, France, 18–21 June 2007, pp. 1–6.

30. T. Piechowiak, Pneumatic train brake simulation method, *Vehicle System Dynamics*, 47(12), 2009, 1473–1492.

31. T. Piechowiak, Verification of pneumatic railway brake models, *Vehicle System Dynamics*, 48(3), 2010, 283–299.

32. M.A. Murtaza, Railway air brake simulation: An empirical approach, *Rail and Rapid Transit*, 207(1), 1993, 51–56.

33. M. Chou, X. Xia, C. Kayser, Modelling and model validation of heavy-haul trains equipped with electronically controlled pneumatic brake systems, *Control Engineering Practice*, 15(4), 2007, 501–509.

34. J. Liu, Y. Liu, Z. Huang, W. Gui, H. Hu, Modelling and control design for an electro-pneumatic braking system in trains with multiple locomotives, *International Journal of Modelling, Identification and Control*, 17(2), 2012, 99–108.

35. T. Vernersson, Temperatures at railway tread braking—Part 1: Modelling, *Rail and Rapid Transit*, 221(2), 2007, 167–182.
36. A.S. Babu, N.S. Prasad, Coupled field finite element analysis of railway block brakes, *Rail and Rapid Transit*, 223(4), 2009, 345–352.
37. B. Ghadimi, F. Kowsary, M. Khorami, Thermal analysis of locomotive wheel-mounted brake disc, *Applied Thermal Engineering*, 51(1–2), 2013, 948–952.
38. Y. Zhu, U. Olofsson, R. Nilsson, A field test study of leaf contamination on railhead surfaces, *Rail and Rapid Transit*, 228(1), 2014, 71–84.
39. M. Malvezzi, L. Pugi, S. Papini, A. Rindi, P. Toni, Identification of a wheel–rail adhesion coefficient from experimental data during braking tests, *Rail and Rapid Transit*, 227(2), 2013, 128–139.
40. S. Teimourimanesh, T. Vernersson, R. Lunden, Modelling of temperatures during railway tread braking: Influence of contact conditions and rail cooling effect, *Rail and Rapid Transit*, 228(1), 2014, 93–109.
41. D.J. Kim, Y.M. Lee, J.S. Park, C.S. Seok, Thermal stress analysis for a disk brake of railway vehicles with consideration of the pressure distribution on a frictional surface, *Materials Science and Engineering A*, 483–484, 2008, 456–459.
42. W.H. Nong, F. Gao, R. Fu, Q.J. Yu, On structure function and the temperature and the thermal stress of brake discs, *Applied Mechanics and Materials*, 80–81, 2011, 521–526.
43. A.E. Beagles, D.I. Fletcher, The aerodynamics of freight: Approaches to save fuel by optimising the utilisation of container trains, *Rail and Rapid Transit*, 227(6), 2013, 635–643.
44. C. Baker, F. Cheli, A. Orellano, N. Paradot, C. Proppe, D. Rocchi, Cross-wind effects on road and rail vehicles, *Vehicle System Dynamics*, 47(8), 2009, 983–1022.
45. C. Baker, H. Hemida, S. Iwnicki, G. Xie, D. Ongaro, Integration of crosswind forces into train dynamic modelling, *Rail and Rapid Transit*, 225(2), 2011, 154–164.
46. F. Cheli, R. Corradi, G. Tomasini, Crosswind action on rail vehicles: A methodology for the estimation of the characteristic wind curves, *Journal of Wind Engineering and Industrial Aerodynamics*, 104–106, 2012, 248–255.
47. P.A. Gaylard, The application of computational fluid dynamics to railway aerodynamics, *Rail and Rapid Transit*, 207(2), 1993, 133–141.
48. C. Biotto, A. Proverbio, O. Ajewole, N.P. Waterson, J. Peiró, On the treatment of transient area variation in 1D discontinuous Galerkin simulations of train-induced pressure waves in tunnels, *International Journal for Numerical Methods in Fluids*, 71(2), 2013, 151–174.
49. C. Baker, The flow around high speed trains, *Journal of Wind Engineering and Industrial Aerodynamics*, 98(6–7), 2010, 277–298.
50. Y.Y. Ko, C.H. Chen, I.T. Hoe, S.T. Wang, Field measurements of aerodynamic pressures in tunnels induced by high speed trains, *Journal of Wind Engineering and Industrial Aerodynamics*, 100(1), 2012, 19–29.
51. O.C. Zienkiewicz, *The Finite Element Method in Engineering Science*, McGraw-Hill, London, UK, 1971.
52. E. Di Gialleonardo, A. Premoli, S. Gallazzi, S. Bruni, Sloshing effects and running safety in railway freight vehicles, *Vehicle System Dynamics*, 51(10), 2013, 1640–1654.
53. P. Wriggers, *Computational Contact Mechanics*, John Wiley & Sons, Chichester, UK, 2002.
54. N.M. Newmark, A method of computation for structural dynamics, *Journal of the Engineering Mechanics Division, ASCE*, 85(3), 1959, 67–94.
55. S.P. Chunduru, M.J. Kim, C. Mirman, Failure analysis of railroad couplers of AAR type E, *Engineering Failure Analysis*, 18(1), 2011, 374–385.
56. C. Baykasoglu, E. Sunbuloglu, S.E. Bozdag, F. Aruk, T. Toprak, A. Mugan, Crash and structural analyses of an aluminium railroad passenger car, *International Journal of Crashworthiness*, 17(5), 2012, 519–528.

57. P.F. Carlbom, Combining MBS with FEM for rail vehicle dynamics analysis, *Multibody System Dynamics*, 6(3), 2001, 291–300.
58. S.W. Kirkpatrick, M. Schroeder, J.W. Simons, Evaluation of passenger rail vehicle crashworthiness, *International Journal of Crashworthiness*, 6(1), 2001, 95–106.
59. H.J. Jang, K.B. Shin, S.H Han, Numerical study on crashworthiness assessment and improvement of composite carbody structures of tilting train using hybrid finite element model, *Advanced Composite Materials*, 21(5–6), 2012, 371–388.
60. Y.Q. Sun, C. Cole, M. Dhanasekar, D.P. Thambiratnam, Modelling and analysis of the crush zone of a typical Australian passenger train, *Vehicle System Dynamics*, 5(7), 2012, 1137–1155.
61. H.J. Cho, J.S. Koo, A numerical study of the derailment caused by collision of a rail vehicle using a virtual testing model, *Vehicle System Dynamics*, 50(1), 2012, 79–108.
62. V. Monfared, A new analytical formulation for contact stress and prediction of crack propagation path in rolling bodies and comparing with finite element model (FEM) results statically, *International Journal of the Physical Sciences*, 6(15), 2011, 3589–3594.
63. J. Cunha, A.G. Correia, Evaluation of a linear elastic 3D FEM to simulate rail track response under a high-speed train, *Proceedings of the Second International Conference on Transportation Geotechnics*, Hokkaido, Japan, 10–12 September 2012, pp. 192–201.
64. Z. Zhou, F. Li, S. Sun, N. Hou, Y. Huang, Mechanical fatigue strength analysis of wheel web for 32.5t axle-load heavy haul wagon, *Proceedings of the Third International Conference on Transportation Engineering*, Chengdu, China, 23–25 July 2011, pp. 2580–2585.
65. E. Di Gialleonardo, F. Braghin, S. Bruni, The influence of track modelling options on the simulation of rail vehicle dynamics, *Journal of Sound and Vibration*, 331(19), 2012, 4246–4258.
66. Y.Q. Sun, M. Dhanasekar, D. Roach, A three-dimensional model for the lateral and vertical dynamics of wagon-track systems, *Rail and Rapid Transit*, 217(1), 2003, 31–45.
67. W. Zhai, Z. Cai, Dynamic interaction between a lumped mass vehicle and a discretely supported continuous rail track, *Computers & Structures*, 63(5), 1997, 987–997.
68. R. Ferrara, G. Leonardi, F. Jourdan, A contact-area model for rail-pads connections in 2-D simulations: Sensitivity analysis of train-induced vibrations, *Vehicle System Dynamics*, 51(9), 2013, 1342–1362.
69. C.J. Greisen, Measurement, simulation, and analysis of the mechanical response of railroad track, Master of Science thesis, University of Nebraska-Lincoln, 2010. Available at: http://digitalcommons.unl.edu/mechengdiss/?utm_source=digitalcommons.unl.edu%2Fmechengdiss%2F9&utm_medium=PDF&utm_campaign=PDFCoverPages.
70. B. Indraratna, H. Khabbaz, W. Salim, D. Christie, Geotechnical properties of ballast and the role of geosynthetics in rail track stabilisation, *Journal of Ground Improvement*, ICE, 10(3), 2006, 91–102, doi: 10.1680/grim.2006.10.3.91.
71. K.H. Chu, V. K. Garg, A. Wiriyachai, Dynamic interaction of railway train and bridges, *Vehicle System Dynamics*, 9(4), 1980, 207–236.
72. G. Diana, F. Cheli, Dynamic interaction of railway systems with large bridges, *Vehicle System Dynamics*, 18(1–3), 1989, 71–106.
73. M.K. Song, H.C. Noh, C.K. Choi, A new three-dimensional finite element analysis model of high-speed train–bridge interactions, *Engineering Structures*, 25(13), 2003, 1611–1626.
74. L. Jamtsho, M. Dhanasekar, Performance testing of a road bridge deck containing flat rail wagons, *Journal of Bridge Engineering*, 18(4), 2013, 308–317.
75. S.P. Jung, T.W. Park, J.H. Lee, Numerical analysis of the dynamic interaction between pantograph and overhead contact line using FEM, *Proceedings of the ASME/ASCE/IEEE Joint Rail Conference*, Pueblo, CO, 16–18 March, 2011, pp. 91–92.

76. S.H. Kia, F. Bartolini, A. Mpanda-Mabwe, R. Ceschi, Pantograph-catenary interaction model comparison, *Proceedings of the 36th Annual Conference on IEEE Industrial Electronics Society*, Glendale, AZ, 7–10 November, 2010, pp. 1584–1589.
77. J. Pombo, J. Ambrósio, M. Pereira, F. Rauter, A. Collins, A. Facchinetti, Influence of the aerodynamic forces on the pantograph–catenary system for high-speed trains, *Vehicle System Dynamics*, 47(11), 2009, 1327–1347.
78. J.H. Lee, T.W. Park, Development of a three-dimensional catenary model using cable elements based on absolute nodal coordinate formulation, *Journal of Mechanical Science and Technology*, 26(12), 2012, 3933–3941.
79. T.J. Park, C.S. Han, J.H. Jang, Dynamic sensitivity analysis for the pantograph of a high-speed rail vehicle, *Journal of Sound and Vibration*, 266(2), 2003, 235–260.
80. H. Bettaieb, Analytical dynamic and quasi-static model of railway vehicle transit to curved track, *Mechanics & Industry*, 13(04), 2012, 231–244.
81. B.G. Eom, H.S. Lee, Assessment of running safety of railway vehicles using multibody dynamics, *International Journal of Precision Engineering and Manufacturing*, 11(2), 2010, 315–320.
82. Y.C. Cheng, C.T. Hsu, Hunting stability and derailment analysis of a car model of a railway vehicle system, *Rail and Rapid Transit*, 226(2), 2012, 187–202.
83. A. Chudzikiewicz, Simulation of rail vehicle dynamics in MATLAB environment, *Vehicle System Dynamics*, 33(2), 2000, 107–119.
84. K.H.A. Abood, R.A. Khan, Railway carriage simulation model to study the influence of vertical secondary suspension stiffness on ride comfort of railway carbody, *Journal of Mechanical Engineering Science*, 225(6), 2011, 1349–1359.
85. M. Arnold, B. Burgermeister, C. Führer, G. Hippmann, G. Rill, Numerical methods in vehicle system dynamics: State of the art and current developments, *Vehicle System Dynamics*, 49(7), 2011, 1159–1207.

5 Multibody Dynamics

5.1 INTRODUCTION TO MULTIBODY DYNAMICS

In any mechanical system there are many interconnected bodies and components which can be considered to be rigid or deformable bodies, or even plastic bodies. Such a system is defined as a multibody system. In general, a multibody system can be a collection of subsystems called bodies, components or substructures. The motions of the subsystems are kinematically constrained because of joints and connections, and each subsystem may undergo large translations and rotational displacements. This book is focussed on the railway system, which basically includes train and track subsystems. A train subsystem, for example, is formed by several locomotives and a series of rail vehicles, longitudinally interconnected by couplers. A rail vehicle generally consists of a vehicle car body, two bogie frames and four wheelsets laterally and vertically interconnected by secondary and primary suspensions.

It is well known that dynamics is concerned with the relationship between motion of bodies and their causes, namely, the forces acting on the bodies and the properties of the bodies (particularly mass and moment of inertia). Rail vehicle dynamics, a branch of multibody dynamics, mainly deals with the vertical and lateral dynamic behaviours of a vehicle's car body and its two bogies during the vehicle moving along a track with geometry irregularities and/or smaller localised defects. The dynamic interactions between a rail vehicle and track can generate great lateral and vertical dynamic forces on wheel–rail contact surfaces, which could cause vehicle derailments or transmit large forces to the vehicle components, causing deterioration and damage. The fundamental theories to establish the relationship between motion of bodies and the forces acting on the bodies in modern dynamics are based on the Newtonian mechanics and its reformulation as Lagrangian mechanics and Hamiltonian mechanics.

The motion of a rigid body in space (or three dimensions) can be completely described by using six degrees of freedom (DOFs) (or generalised coordinates)—three of them translational degrees of freedom and three rotational degrees of freedom. The degrees of freedom denote the number of independent kinematical possibilities to move. In other words, DOFs are the minimum number of parameters required to completely define the position of an entity in space. In the case of planar motion (or two-dimensional motion), a body has only three DOFs, one rotational and two translational.

Nowadays, multibody dynamics is closely related to many fields of engineering research, especially in rail train, rail vehicle and rail vehicle—track interaction dynamics. As an important feature, multibody dynamics has offered an algorithmic, computer-aided way to model, analyse, simulate and optimise the motion of possibly thousands of interconnected bodies.

5.2 KINEMATICS

Kinematics is the branch of mechanics that describes the motion of bodies without consideration of the causes of motion. To describe motion of a rigid body, kinematics studies the trajectory of the body and its differential properties such as velocity and acceleration. As stated in Section 5.1, the motion of a rigid body in space can be described by using six DOFs—three translations and three rotations of the body. For simplicity, the two-dimensional motion of a rigid body is illustrated. Figure 5.1 shows a rigid body (vehicle car body) denoted as body i in a two-dimensional plane (XY).

Let XY be a fixed coordinate system (or global coordinate system) and X_iY_i be a moving body coordinate system (or body coordinate system) whose origin O_i (x_{io}, y_{io}) is fixed to a point on the rigid body. The vector $\mathbf{d}_i = [d_{ix}\ d_{iy}]^T$ describes the translation of the origin of the body coordinate system while the angle θ_{iz} describes the rotation of the body about the Z axis passing through the origin point. According to Chasles' theorem, the most general motion of a rigid body is equivalent to a translation of a point on the body plus a rotation about an axis passing through that point. Therefore, the set of Cartesian coordinates q_i can be defined as

$$q_i = [d_{ix} d_{iy} \theta_{iz}]^T \tag{5.1}$$

Equation 5.1 can be used to determine the body motion in a two-dimensional plane (XY). Hence, the position, velocity and acceleration of an arbitrary point on a rigid body can be described in terms of these coordinates.

The global position of any point p_i on the body (shown in Figure 5.1) can be defined as

$$\mathbf{D}_i = \mathbf{d}_i + \mathbf{u}_i \tag{5.2}$$

where $\mathbf{D}_i = [D_{ix}\ D_{iy}]^T$ is the global position vector of point p_i, $\mathbf{d}_i = [d_{ix}\ d_{iy}]^T$ is the global position vector of the origin O_i of the body coordinate system, and $\mathbf{u}_i = [u_{ix}\ u_{iy}]^T$ is the position vector of point p_i with respect to O_i, which can be rewritten as

$$\mathbf{u}_i = x_{ip}\boldsymbol{i}_i + y_{ip}\boldsymbol{j}_i \tag{5.3}$$

where $\boldsymbol{i}_i$ and $\boldsymbol{j}_i$ are unit vectors along the body axes X_i and Y_i respectively; x_{ip} and y_{ip} are the coordinates of point p_i in the body coordinate system and are constant

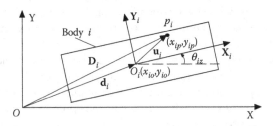

FIGURE 5.1 Rigid-body kinematics.

because the body is assumed to be rigid. For the determination of the velocity vector of point p_i, Equation 5.2 is differentiated with respect to time. This yields

$$\mathbf{v}_i = \frac{d\mathbf{D}_i}{dt} = \dot{\mathbf{d}}_i + \dot{\mathbf{u}}_i = \dot{\mathbf{d}}_i + x_{ip}\frac{d\boldsymbol{i}_i}{dt} + y_{ip}\frac{d\boldsymbol{j}_i}{dt} = \dot{\mathbf{d}}_i + x_{ip}\dot{\theta}_{iy}\boldsymbol{j}_i - y_{ip}\dot{\theta}_{iy}\boldsymbol{i}_i \quad (5.4)$$

An angular velocity vector $\boldsymbol{\omega}_i$ of body i is introduced as

$$\boldsymbol{\omega}_i = \dot{\theta}_{iz}\mathbf{k}_i \tag{5.5}$$

where $\mathbf{k}_i$ is a unit vector along the body axis $\mathbf{Z}_i$ that passes through point O_i and is perpendicular to $\boldsymbol{i}_i$ and $\boldsymbol{j}_i$. Therefore, the following equation can be derived:

$$\boldsymbol{\omega}_i \times \mathbf{u}_i = \begin{vmatrix} 0 & \dot{\theta}_{iz} \\ y_{ip} & 0 \end{vmatrix} \boldsymbol{i}_i - \begin{vmatrix} 0 & \dot{\theta}_{iz} \\ x_{ip} & 0 \end{vmatrix} \boldsymbol{j}_i + \begin{vmatrix} 0 & 0 \\ x_{ip} & y_{ip} \end{vmatrix} \mathbf{k}_i = x_{ip}\dot{\theta}_{iy}\boldsymbol{j}_i - y_{ip}\dot{\theta}_{iy}\boldsymbol{i}_i \quad (5.6)$$

Equation 5.4 can be rewritten as

$$\mathbf{v}_i = \dot{\mathbf{d}}_i + \dot{\mathbf{u}}_i = \dot{\mathbf{d}}_i + \boldsymbol{\omega}_i \times \mathbf{u}_i \tag{5.7}$$

By differentiating Equation 5.7 with respect to time, an expression for the acceleration vector can be obtained as

$$\mathbf{a}_i = \frac{d\mathbf{v}_i}{dt} = \ddot{\mathbf{d}}_i + \ddot{\mathbf{u}}_i = \ddot{\mathbf{d}}_i + \dot{\boldsymbol{\omega}}_i \times \mathbf{u}_i + \boldsymbol{\omega}_i \times \dot{\mathbf{u}}_i \tag{5.8}$$

If an angular acceleration vector, $\boldsymbol{\alpha}_i$, of body i is introduced as

$$\boldsymbol{\alpha}_i = \ddot{\theta}_{iz}\mathbf{k}_i \tag{5.9}$$

Equation 5.8 can be written as

$$\mathbf{a}_i = \ddot{\mathbf{d}}_i + \boldsymbol{\alpha}_i \times \mathbf{u}_i + \boldsymbol{\omega}_i \times (\boldsymbol{\omega}_i \times \mathbf{u}_i) \tag{5.10}$$

In Equation 5.10, $\ddot{\mathbf{d}}_i$ is the acceleration of the origin of the coordinate system. The term $\boldsymbol{\alpha}_i \times \mathbf{u}_i$ is called the tangential component of the acceleration of point p_i with respect to O_i because its direction is perpendicular to both vectors $\boldsymbol{\alpha}_i$ and $\mathbf{u}_i$. The term $\boldsymbol{\omega}_i \times (\boldsymbol{\omega}_i \times \mathbf{u}_i)$ is the normal component of the acceleration of point p_i with respect to O_i because it is directed from p_i to O_i.

It is noted that the velocity expressed in Equation 5.7 and the acceleration in Equation 5.10 can be written in terms of the coordinates $q_i = [d_{ix}d_{iy}\theta_{iz}]^T$ and their time derivatives. In the spatial (or three-dimensional) analysis, similar expressions for the velocity and acceleration of any point on a rigid body can be obtained.

5.3 DYNAMICS

Dynamics is a branch of classical mechanics concerned with the study of forces and torques and their effect on motion. In this section, techniques for developing the dynamic equations of motion of multibody systems consisting of interconnected rigid bodies are briefly introduced [1–3].

5.3.1 NEWTON–EULER EQUATIONS

It is known that the unconstrained three-dimensional motion of the rigid body can be described using six equations—three translational equations of the rigid body and three rotational equations are associated with the body. It is important for the origin O_i of the body coordinate system to be fixed to the body mass centre, which can significantly simplify the dynamic equations. In this circumstance, the translational equations are called the Newton equations while the rotational equations are called the Euler equations. Following the case of planar motion in the previous section, the Newton–Euler equations can be written, for body i in a multibody system, as

$$\begin{cases} m_i \mathbf{a}_i = \mathbf{F}_i \\ J_i \ddot{\theta}_{iz} = \mathbf{M}_i \end{cases} \tag{5.11}$$

where m_i is the total mass of the rigid body, $\mathbf{a}_i$ is a two-dimensional vector that defines the absolute acceleration of the body mass centre, $\mathbf{F}_i$ is the vector of forces acting on the body mass centre, J_i is the mass moment of inertia defined with respect to the mass centre, and $\mathbf{M}_i$ is the moment acting on the body.

5.3.2 D'ALEMBERT'S PRINCIPLE AND GENERALISED FORCES

Generally, a multibody system with n coordinates and n_c constraint equations which are linearly independent has $(n - n_c)$ independent coordinates, also called the system degrees of freedom (DOFs). Generalised coordinates ($\mathbf{q}_j$, $j = 1,2,...,m \leq (n - n_c)$) are defined as the minimum number of independent coordinates that define the configuration of a system. A constraint is actually represented by the force that prevents a body moving in a certain path. A more efficient approach used to generate the dynamic equations of motion of a multibody system, incorporating the constraints on the system, might be D'Alembert's principle because the kinetic and potential energies of a multibody system are much easier to write down and calculate than the forces since energy is a scalar while forces are vectors. The virtual work is introduced in this principle to study the dynamic equilibrium of a multibody system.

The body position vector in a multibody system $\mathbf{r}$ in a standard coordinate system (Cartesian, spherical etc.) is related to the generalised coordinates by transformation equations given by

$$\mathbf{r}_i = \mathbf{r}_i\left(\mathbf{q}_1, \mathbf{q}_2, \ldots, \mathbf{q}_m, t\right), \quad i = 1, 2, \ldots, n \tag{5.12}$$

The virtual work of forces acting on a multibody system is obtained from the scalar product of each force with the virtual displacement of its point of application, expressed as

$$\delta W = \sum_{i=1}^{n} \mathbf{F}_i \cdot \delta \mathbf{r}_i \tag{5.13}$$

From Equation 5.12, the virtual displacements $\delta \mathbf{r}_i$ can be given by

$$\delta \mathbf{r}_i = \sum_{j=1}^{m} \frac{\partial \mathbf{r}_i}{\partial \mathbf{q}_j} \cdot \delta \mathbf{q}_j \tag{5.14}$$

The virtual work of forces in the system in terms of the generalised coordinates becomes

$$\delta W = \sum_{j=1}^{m} \sum_{i=1}^{n} \mathbf{F}_i \cdot \frac{\partial \mathbf{r}_i}{\partial \mathbf{q}_j} \cdot \delta \mathbf{q}_j$$

The generalised forces can be defined as

$$Q_j = \sum_{i=1}^{n} \mathbf{F}_i \cdot \frac{\partial \mathbf{r}_i}{\partial \mathbf{q}_j}$$

Q_j is known as the generalised force associated with the virtual displacement $\delta \mathbf{q}_j$. Based on Newton's equations, Equation 5.13 can be rewritten as

$$\delta W = \sum_{i=1}^{n} \left(\mathbf{F}_i - m_i \mathbf{a}_i \right) \cdot \delta \mathbf{r}_i = 0 \tag{5.15}$$

The above dynamic equilibrium of a multibody system represents D'Alembert's principle.

5.3.3 LAGRANGE'S EQUATION

It is known that the kinetic energy T for the system can be defined by

$$T = \frac{1}{2} \sum_{i=1}^{n} m_i \dot{\mathbf{r}}_i \cdot \dot{\mathbf{r}}_i \tag{5.16}$$

Equation 5.16 is partially differentiated with respect to the generalised coordinates $\mathbf{q}_j$ and generalised velocities $\dot{\mathbf{q}}_j$, and the following equations are obtained, respectively:

$$\frac{\partial T}{\partial \mathbf{q}_j} = \sum_{i=1}^{n} m_i \dot{\mathbf{r}}_i \cdot \frac{\partial \dot{\mathbf{r}}_i}{\partial \mathbf{q}_j} \quad \text{and} \quad \frac{\partial T}{\partial \dot{\mathbf{q}}_j} = \sum_{i=1}^{n} m_i \dot{\mathbf{r}}_i \cdot \frac{\partial \dot{\mathbf{r}}_i}{\partial \dot{\mathbf{q}}_j} \tag{5.17}$$

Because $(\partial \dot{\mathbf{r}}_i / \partial \dot{\mathbf{q}}_j) = (\partial \mathbf{r}_i / \partial \mathbf{q}_j)$, so,

$$\frac{\partial T}{\partial \dot{\mathbf{q}}_j} = \sum_{i=1}^{n} m_i \dot{\mathbf{r}}_i \cdot \frac{\partial \mathbf{r}_i}{\partial \mathbf{q}_j} \tag{5.18}$$

Equation 5.18 is differentiated with respect to time and yields:

$$\frac{d}{dt}\frac{\partial T}{\partial \dot{\mathbf{q}}_j} = \sum_{i=1}^{n} m_i \ddot{\mathbf{r}}_i \cdot \frac{\partial \mathbf{r}_i}{\partial \mathbf{q}_j} + \sum_{i=1}^{n} m_i \dot{\mathbf{r}}_i \cdot \frac{\partial \dot{\mathbf{r}}_i}{\partial \mathbf{q}_j} = Q_j + \frac{\partial T}{\partial \mathbf{q}_j}$$

Finally, the above equation results in

$$Q_j = \frac{d}{dt}\left(\frac{\partial T}{\partial \dot{\mathbf{q}}_j}\right) - \frac{\partial T}{\partial \mathbf{q}_j} \tag{5.19}$$

This is an important Lagrange equation based on the dynamic equilibrium of a multibody system with the virtual work and generalised coordinates. However, the kinematic constraint equations may exist because of connections between bodies or specified motion trajectories in a multibody system. The constraints are sometimes considered as the classical constraints which are usually a set of algebraic equations that define the relative translations or rotations between bodies. In a nonholonomic system, there are possibilities to constrain the relative velocities between bodies. In addition, there are non-classical constraints that might even introduce a new unknown coordinate. Therefore, the equation expressed in Equation 5.19 cannot actually be used in building a dynamics model.

5.3.4 DYNAMIC EQUATIONS

A technique called the augmented formulation [1] can be applied to formulate the dynamic equations of constrained multibody systems.

The constraint equations of a multibody system can be written as

$$\mathbf{C}(\mathbf{q}, t) = 0 \tag{5.20}$$

where $\mathbf{C} = [C_1(\mathbf{q},t)C_2(\mathbf{q},t)\cdots C_{n_c}(\mathbf{q},t)]^{\mathrm{T}}$ is the vector of constraint equations; $\mathbf{q}$ is the vector of generalised coordinates; and n_c is the number of constraint equations. For a virtual displacement $\delta\mathbf{q}$, Equation 5.20 becomes

$$\mathbf{C}_q \delta \mathbf{q} = 0 \qquad (5.21)$$

where $\mathbf{C}_q$ is the constraint Jacobian matrix.

In the augmented formulation [1], the Lagrange multipliers can be used for both holonomic and nonholonomic systems. Provided that the constraint relationships are velocity-dependent and non-integrable, the following equation exists:

$$\left(\mathbf{C}_q \delta \mathbf{q} \right)^{\mathrm{T}} \boldsymbol{\lambda} = 0 \qquad (5.22)$$

where $\boldsymbol{\lambda} = [\lambda_1 \ \lambda_2 \ \dots \ \lambda_{n_c}]^{\mathrm{T}}$ is the vector of Lagrange multipliers.

Based on the principle of virtual work,

$$\delta W = \left(\mathbf{M} \ddot{\mathbf{q}} - \mathbf{Q} \right)^{\mathrm{T}} \cdot \delta \mathbf{q} = 0 \qquad (5.23)$$

where $\mathbf{M}$ is the system mass matrix, $\mathbf{Q} = \mathbf{Q}_v + \mathbf{Q}_e$ is the total vector of forces ($\mathbf{Q}_v$ is the vector of centrifugal and inertia forces and $\mathbf{Q}_e$ is the vector of externally applied forces including gravity, spring, damper, and actuator forces).

Equations 5.22 and 5.23 can be combined to yield:

$$\delta \mathbf{q}^{\mathrm{T}} \left(\mathbf{M} \ddot{\mathbf{q}} - \mathbf{Q} + \mathbf{C}_q^{\mathrm{T}} \right) = 0 \qquad (5.24)$$

By partitioning the coordinates as dependent and independent, $\mathbf{M}$ and $\mathbf{Q}$ can be written as

$$\mathbf{M} = \begin{bmatrix} \mathbf{M}_{dd} & \mathbf{M}_{di} \\ \mathbf{M}_{id} & \mathbf{M}_{ii} \end{bmatrix} \quad \text{and} \quad \mathbf{Q} = \begin{bmatrix} \mathbf{Q}_d \\ \mathbf{Q}_i \end{bmatrix}$$

where subscripts $\mathbf{d}$ and $\mathbf{i}$ represent dependent and independent, respectively. The components of the virtual displacement vector $\delta \mathbf{q}$ in Equation 5.23 are not independent because of the holonomic or nonholonomic constraint equations. It is supposed that λ_k ($k = 1, 2, \dots, n_c$) is selected, so that

$$\mathbf{M}_{dd} \ddot{\mathbf{q}}_d + \mathbf{M}_{di} \ddot{\mathbf{q}}_i - \mathbf{Q}_d + \mathbf{C}_{q_d}^{\mathrm{T}} \boldsymbol{\lambda} = 0 \qquad (5.25)$$

where $\mathbf{q}_d = [q_1 \ q_2 \ \dots \ q_{n_c}]^{\mathrm{T}}$ are the dependent variables. Substituting Equation 5.25 into Equation 5.24 gives

$$\delta \mathbf{q}_i^{\mathrm{T}} \left(\mathbf{M}_{ii} \ddot{\mathbf{q}}_i + \mathbf{M}_{id} \ddot{\mathbf{q}}_d - \mathbf{Q}_i + \mathbf{C}_{q_i}^{\mathrm{T}} \boldsymbol{\lambda} \right) = 0 \qquad (5.26)$$

Since the elements of $\delta \mathbf{q}_i$ in this equation are independent, so,

$$\mathbf{M}_{ii} \ddot{\mathbf{q}}_i + \mathbf{M}_{id} \ddot{\mathbf{q}}_d - \mathbf{Q}_i + \mathbf{C}_{q_i}^{\mathrm{T}} \boldsymbol{\lambda} = 0 \qquad (5.27)$$

Because $\mathbf{q_d}$ and $\mathbf{q_i}$ are the partitions of $\mathbf{q}$, one equation can be obtained by combining Equations 5.25 and 5.27 to yield

$$\mathbf{M\ddot{q}} - \mathbf{Q} + \mathbf{C_q^T}\boldsymbol{\lambda} = 0 \qquad (5.28)$$

The above equation represents a set of differential equations of motion which, along with the constraint equations, can be solved for the vector of system generalised coordinates $\mathbf{q}$ and the vector of the Lagrange multipliers $\boldsymbol{\lambda}$. This equation can be used to develop the dynamic equilibrium equations of motion for the dynamic analysis of a multibody system subject to both holonomic and nonholonomic constraints.

5.4 ELEMENTS

The previous sections describe the theoretical background on how to develop the dynamic equilibrium equations of a multibody system model and to further analyse its dynamic behaviours. Taking the establishment of a rail vehicle model as an example, this section will present the elements used to construct such a model. The elements can be divided into three main categories—inertial elements, suspension elements and constraint elements. The definitions and descriptions of the elements in VAMPIRE software are described in this section.

5.4.1 Rail Vehicle Modelling Axis System

A rail vehicle model is a collection of inertial elements (masses and wheelsets) connected by suspension elements and/or constraint elements. It may, as required, represent a single vehicle or a train of multiple vehicles. The model requires an associated axis system as shown in Figure 5.2.

Positions of inertial elements and suspension elements are specified in terms of longitudinal position, lateral position and height above rail. The longitudinal axis (X) is usually fixed at the mid-point of the vehicle or train; positive longitudinal positions are in the direction of travel. The lateral axis (Y) is relative to the track centreline. The vertical axis (Z) is relative to the height above the rail which is normally considered positive, but some commercial software packages for rail vehicle dynamics simulation use downwards as positive.

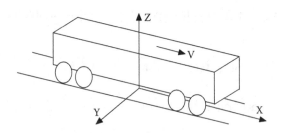

FIGURE 5.2 Vehicle modelling axis system.

5.4.2 Inertia Elements

The inertia (or mass) element is the basic building block of a vehicle model, and is usually used to represent vehicle body, bogie frames, wheelsets, traction motors, rail and track blocks and so forth. A mass element has up to six degrees of freedom, representing longitudinal, lateral and vertical translations and roll, pitch and yaw rotations. When defining a mass element in a vehicle model, the required properties are its mass and the roll, pitch and yaw inertial moments. The location of the centre of gravity of a mass element must also be supplied as its longitudinal distance, lateral distance, and height based on the model global or local coordinate system.

Sometimes, a wheelset is considered as a special mass element because it is a rolling mass and the pitch degree of freedom of a wheelset is a function of the forward velocity of the wheelset. Suspension elements may not be allowed to attach to the wheelset pitch freedom, but external forcing may be applied to the wheelset pitch freedom.

5.4.3 Suspension Elements

A multibody system consists of many interconnected bodies. The main interconnection elements in rail vehicle modelling are called the suspension elements, which connect the masses, wheelsets and even flexible modes in a vehicle model. The suspension elements include the stiffness, bumpstop, damper, friction, pinlink, shear spring, airspring, bush and constraint elements.

The line and rotational elements consist of the stiffness, bumpstop, damper and friction elements. The common characteristics of these elements are that they have a fixed line of action acting along the element's axis, which is determined by the positions of its ends.

The stiffness element represents a single linear stiffness value, as shown in Figure 5.3.

The force **F** across the element is calculated as

$$\mathbf{F} = \mathrm{K} \cdot \mathbf{d}$$

where **d** is the deflection across the element and K is the stiffness coefficient.

The bumpstop element represents a non-linear stiffness specified as a force–displacement characteristic as shown in Figure 5.4.

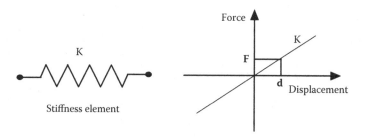

FIGURE 5.3   Stiffness element.

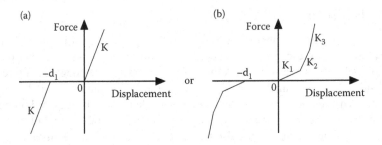

FIGURE 5.4 Bumpstop element. (a) Constant rate, (b) variable rate.

The force **F** across the element is calculated for the constant rate, non-linear characteristic shown in Figure 5.4a as

$$\mathbf{F} = \begin{cases} 0 & -d_1 \le \mathbf{d} \le 0 \\ K \cdot \mathbf{d} & \mathbf{d} < -d_1 \text{ and } \mathbf{d} > 0 \end{cases}$$

The damper represents a viscous damping element with either a constant rate or a variable rate specified as a force–velocity characteristic, as shown in Figure 5.5. The damper also includes a stiffness element in series to represent the flexibility of the mountings.

The force **F** across the element with a constant rate is calculated as

$$\mathbf{F} = C \cdot \mathbf{v}$$

where **v** is the velocity across the element and C is the constant damping coefficient.

The friction element can have a simple break out force (its characteristic is shown in Figure 5.6) or a load-dependent breakout where the load may be determined from a spring, bumpstop, pinlink or shear element. Friction elements always include a series stiffness, which has a damper in parallel to avoid ringing. Two friction elements in perpendicular directions may be linked so that the breakout occurs as a vector sum of the forces in the two directions, to represent the full effect of friction due to the contact of two planar surfaces. Based on Figure 5.6, the break out force can be expressed as

$$\mathbf{F} = -\mu \mathbf{F}_n \cdot \text{sign}(\mathbf{v})$$

where μ is the friction coefficient.

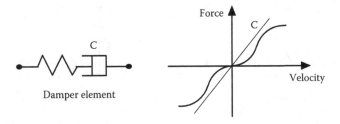

FIGURE 5.5 Damper element.

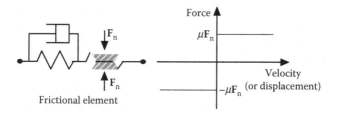

FIGURE 5.6   Friction element.

The pinlink element is a non-linear element. Pinlink elements have been especially developed for applications in the following situations:

- The shape of a body in a multibody system is slender (e.g. a traction rod, swing link, viscous damper and the like)—the length of the pinlink element is a very important property as it affects its dynamic behaviour;
- Large displacement effects across the suspension elements are expected.

It is noted that the pinlink element is ball jointed at each end and so cannot generate any moments. The only internal forces in the element act along its axis. This element can model a variety of suspension components as it can possess a combination of stiffness, damping and friction properties as shown in Figure 5.7.

Shear elements are developed in order to support a significant static load only in the vertical direction and meanwhile to provide flexibility in tangent directions perpendicular to this static load. The secondary suspensions with airsprings (fitted on most modern passenger vehicles) or flexicoil springs (modern freight vehicles) and the primary suspensions with coil or rubber springs can be the most accurately modelled using this type of element. A simple model would represent this situation by two separate springs, one acting vertically and the other laterally.

Airspring elements are designed to represent a detailed model of pneumatic damping in the vertical direction, and a non-uniform stiffness distribution in the lateral and longitudinal directions, giving different moments at each end, thus allowing it to represent the characteristics of airsprings. It can also have a non-linear stiffness characteristic, and it offers a means of representing the frequency-dependent stiffness and hysteresis effects of rubber components. The vertical and lateral behaviours

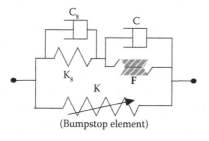

(Bumpstop element)

FIGURE 5.7   Pinlink element.

of airspring elements are independent. The vertical behaviour depends on the stiffness and damping, both of which are related to the frequency and load. However, the determination of the lateral forces and moments depends on the vertical load. In many vehicles, the damping provided by airsprings is the only vertical damping in the secondary suspension.

Shear spring elements are designed to have uniform stiffness properties so as to act symmetrically from top to bottom. They are appropriate for modelling suspension elements such as flexicoil springs, or uniform rubber stack suspension elements, where large vertical loads are carried with lateral flexibility.

The bush element is equivalent to linear springs with dampers in parallel acting in six directions at a single point, which is intended to represent a rubber bush or similar suspension element. It can also be used for elements such as anti-roll bars, in which case it not only provides roll stiffness, but can also represent the parasitic stiffness and damping in other directions.

5.4.4 CONSTRAINT ELEMENTS

The constraints have been treated in different ways in the linear and non-linear analysis programs. In the linear eigenvalue and response programs, the generalised constraints allow two bodies to be rigidly connected together for translation in a particular direction and/or rotation about a particular axis at a specified point in space. The programs cope with this by eliminating one constraint and applying an alternative equivalent constraint.

In the non-linear analysis, based on the content of Section 5.3.4, if more accurate constraint equations could be written, they are time varying. Multiplication of the system equations by the constraint matrix would be required at every time step. The calculation time would be unacceptable. In fact, a set of constraints at a point in the vehicle is replaced by a bush element but with the stiffness, damping and series stiffness for each direction.

5.5 RIGID BODY VERSUS FLEXIBLE BODY

Why a body in a multibody system is defined as either a rigid body or a flexible/deformable/plastic body depends upon the considerations regarding this body. If the deformation of the body under analysis is considered to be so small that the body deformation has no effect on the gross body motion, this body can be treated as a rigid body; otherwise, it must be treated as a flexible or deformable body. Another consideration may be the frequency range required.

In rail vehicle modelling, for example, if the critical frequency range for a rail vehicle's dynamic behaviour is below 10 Hz, all bodies making up the vehicle including the car body, two bogie frames and the wheelsets can be considered as rigid bodies. This would apply to a vehicle being modelled for lateral hunting stability analysis because the hunting frequency is much less than 10 Hz. However, in an investigation of ride comfort for a rail passenger vehicle, it is better to consider the car body as deformable because the natural frequencies of basic vibrating modes of a car body are normally just above 10 Hz.

It is known that the choice of the centre of mass as the origin of the body coordinate system leads to significant simplifications in the form of the dynamic equations. As the result of such a choice of the body reference system, the Newton–Euler equations have no inertia coupling between the translational and rotational coordinates of the rigid body. However, such a decoupling of the coordinates becomes more difficult when deformable bodies are considered. In fact, for a deformable body, two arbitrary points on it move relative to each other, and consequently the body coordinates are no longer sufficient to describe the kinematics of deformable bodies. An infinite number of coordinates are actually required to define the exact position of each point on the deformable body, and the exact modelling of the dynamics of deformable bodies demands an infinite number of degrees of freedom. To avoid the computational difficulties encountered in dealing with infinite dimensional spaces, approximation techniques such as the Rayleigh–Ritz method and finite-element methods are used to reduce the number of coordinates to a finite set.

In the Rayleigh–Ritz method, the shape of deformation of the body is approximated using a finite set of known functions that define the body deformation with respect to its reference system. The dynamics of the deformable body can be modelled using a finite set of elastic coordinates. When the Rayleigh–Ritz method meets the difficulty of determining these approximation functions, or the deformable bodies have complex geometrical shapes, this problem can be solved by using the finite element methods. In the finite element methods, the deformable bodies are discretised into smaller various elements, including truss, beam, rectangular and triangular elements used in the planar analysis, and beam, plate, solid, tetrahedral and shell elements used in the three-dimensional analysis as required depending on the shapes of bodies and the calculation requirements. The interpolating polynomials and the nodal coordinates can define the assumed displacement field of the finite element in terms of an element shape function.

In detailed track modelling, for example, the Rayleigh–Ritz method offers an approximate solution for this problem. In this method, the two rails in track dynamics modelling are required to be considered as flexible bodies using beam theory, for example, Timoshenko beam theory, and their dynamic equilibrium equations are solved using the Rayleigh–Ritz method.

5.6 MULTIBODY DYNAMICS SOFTWARE FOR RAIL VEHICLE DYNAMICS

Dynamic simulations of rail vehicle dynamic behaviours due to wheel–rail interactions can be performed using a computer program or software package. The modelling of rail vehicle dynamics is typically described by a series of ordinary differential equations or partial differential equations. As such, a mathematical model incorporates complex wheel–rail contacts and real-world suspension elements such as friction, bumpstops and the like, and the equations become nonlinear. This requires numerical methods to solve the equations. A numerical simulation is done by stepping through a time interval and calculating the integral of the derivatives by approximating the area under the derivative curves. Some methods use a fixed step through the interval, and others use an adaptive step that can shrink or grow automatically to maintain an acceptable error tolerance.

Dynamic simulations of rail vehicle dynamic behaviours have provided a lot of benefits to the rail industry and its regulators. Firstly, models of dynamic simulation in rail vehicle dynamics can be run to give a virtual dynamic response close to the actual system. This is very useful and important in vehicle derailment investigation and prevention as well as vehicle lateral hunting stability analysis. Secondly, some rail vehicle acceptance procedures or standards have now allowed dynamic simulations to be substituted for field tests, saving significant costs. Finally, the dynamic simulations of rail vehicle dynamic behaviours are necessary for a rail vehicle design, especially for suspension element designs. Parametric sensitivity studies utilising the simulations can ensure a rail vehicle design reaches an optimum outcome.

The following subsections will briefly introduce some widely used commercial software packages for dynamic simulations of rail vehicle dynamic behaviours.

5.6.1   NUCARS

NUCARS, developed by the Transportation Technology Centre Inc., a wholly owned subsidiary of the Association of American Railroads, has been widely adopted and has become an industry standard [4]. NUCARS, as a computer simulation model of a general multibody rail vehicle dynamics system, can perform a variety of roles:

- To simulate and predict the dynamic response of any rail vehicle (including locomotives and passenger, transit or freight vehicles) to specified track conditions or on any type of track geometry including special track work such as turnouts and guardrails;
- To evaluate and compare new vehicle designs as well as to perform failure analyses such as derailment studies and dynamic stability analysis;
- To investigate and improve the ride quality.

NUCARS is designed to simulate the dynamic interaction of any rail vehicle and any track. Any number of bodies, degrees of freedom, and suspension and connection elements to describe a rail vehicle and track system can be selected by the user. Therefore, it can be used to study the dynamic interactions of rail vehicles and tracks, to predict a rail vehicle's stability, ride quality, vertical and lateral dynamics, and steady-state and dynamic curving performances. The detailed nonlinear modelling of wheel–rail contact based on Kalker's complete non-linear creep theory is included in the model. Simulations of any type of freight, passenger, transit and locomotive rail vehicles can be undertaken while simulations of any type of track can include hypothetical track geometries or measured track supplied by the user, including turnouts and guard rails.

The general and special applications of NUCARS can be included in the following aspects:

- *General Areas:*
  - Vehicle Design Optimisation: The effects on rail vehicle ride quality, vehicle curving and safety performance (L/V ratios and wheel unloading

rates) can be easily simulated and determined using a NUCARS model to evaluate changes to suspension characteristics and vehicle geometry. There are a wide range of suspension elements incorporated in NUCARS, including a number of detailed friction models designed to simulate specific railroad vehicle suspension components, as well as normal spring and damping components. Also, there are options for advanced vehicle designs to include items such as articulation, self-steering axles, independently rotating wheels, air bag suspensions and viscous dampers. Special connection elements are included to represent tilting and active suspensions.

- Track Component Design: A NUCARS model can be used to explore changes in track design such as turnout geometry and check (guard) rail configuration. The wheel–rail interaction forces can be calculated and track designs optimised to minimise their effects on the track structure.

- Vehicle Certification: A NUCARS model can be an especially useful tool in assessing vehicle safety performance for issuing vehicle certification. NUCARS can significantly improve the certification tests, which various rail authorities specify for new freight vehicles, by permitting simulations and predictions of actual tests as well as conditions which it may not be possible to test. NUCARS simulations can be conducted up to the point of derailment to establish overall windows of performance.

- *Special Areas:*
  - Derailment Investigation: Derailment conditions can be simulated with the input of measured track geometry data, and actual wheel and rail profile shapes. The specifications of actual vehicle suspension conditions and the presence of unusual conditions and external forces acting on the vehicle can be input to the NUCARS model to calculate the wheel–rail interaction forces and other dynamic conditions that lead to derailment.

  - Vehicle Ride Quality: Passenger comfort evaluations can be performed with a NUCARS model by simulating vehicles running over measured track data. The simulations of actual transducers (e.g. accelerometers) mounted anywhere on the vehicle body can be used for the ride quality investigations.

  - Wheel and Rail Profile Design Optimisation: The effects of wheel and rail profiles on wheel–rail forces, wheel–rail wear and vehicle dynamic performance can be explored by using NUCARS wheel–rail interaction calculations. The trade off between high conicity wheel profiles for good curving performance and low conicity profiles for high-speed stability can be quickly quantified. Output of various wheel–rail contact parameters such as creep forces, contact angles and wheel rolling radii allow a detailed study of the wheel–rail interaction.

  - Wheel–Rail Lubrication Studies: The analyses of the effects of lubrication on wheel and rail wear, rolling resistance and vehicle curving performance can be made by varying the coefficient of friction in the tread, flange and back of the flange regions.

- Rolling Contact Fatigue (RCF) Studies: The significant influence on RCF can be assessed for small variations of wheel and rail profiles, wheel–rail lubrication, all in combination with vehicle curving dynamics. NUCARS has the ability to evaluate all these parameters singly or in combination, allowing detailed studies of the RCF mechanisms.
- Dynamic Clearance Envelope Calculations: A NUCARS model can be used to calculate the clearances between moving vehicles on adjacent tracks, and between vehicles and stationary objects such as bridges and platforms based on the simulations of vehicles running over various radii curves, superelevation and track irregularities.

5.6.2   GENSYS

GENSYS is a software tool for modelling rail vehicles running on tracks [5], but in its design GENSYS is a general multi-purpose software package for modelling mechanical, electrical and/or multibody systems. Modelling of rail vehicles using computers started in 1971 in Sweden, and the program was firstly named LSTAB with linear modelling in the frequency domain comprising a bogie with two wheelsets. In 1973, the development of a non-linear time-domain simulation program began, in which a whole rail vehicle could be modelled. The program consisted of the following two parts: SIMFO-L taking lateral, roll and yaw motions into account and SIMFO-V taking longitudinal, vertical and pitch motions into account. In 1992, a three-dimensional general multibody-dynamic program was developed and was given the name GENSYS. At the same time, the development of the software package moved into a new company, AB DEsolver, which has the sole task of developing and supporting the package.

The main calculation programs in GENSYS are listed in Table 5.1

All the four major calculation programs are very general in their basic design, and with the GENSYS input data syntax it is easy to create the models of systems. If a sub-system is written in an m-file for MATLAB® or Octave, it is possible to make a co-simulation with the cosim_server command. The coupling between wheel and rail can be modelled in many ways.

To simplify the generation of input data, the GENSYS package consists of the pre-processors listed in Table 5.2.

TABLE 5.1
Main Programs in GENSYS

QUASI	Quasistatic analysis
MODAL	Modal analysis
FRESP	Frequency-response analysis
TSIM	Time-domain integration

TABLE 5.2

Pre-Processors in GENSYS

TRACK	Generation of track irregularity files
KPF	Generation of wheel–rail geometrical properties
MISC	Miscellaneous programs for vehicle and track property input etc.
NPICK	Adding flexible modes to rigid bodies
OPTI	Runs sequences of calculations

As indicated in Table 5.2, the GENSYS package can simulate a system combining multi-flexible bodies and rigid bodies.

GENSYS includes three powerful postprocessors, GLPLOT, GPLOT and MPLOT. GPLOT is a 3-D visualisation and animation program which can also animate the wheel–rail-contact conditions. GLPLOT is similar to GPLOT but is written in OpenGL to give an improved appearance. MPLOT is used for post-processing and plotting of the results. Plotting can be 2- or 3-D, consisting of curves and/or scalars. Post-processing in MPLOT consists of algebraic operations, filtering operations, Fourier transform operations, different ride comfort assessments and various statistical evaluation methods.

The main applications for rail vehicle dynamic analysis using GENSYS include:

- Wheel unloading on track twists;
- Carbody roll coefficient;
- Critical speed;
- Pantograph sway;
- Maximum track shift forces;
- Maximum flange climb ratio;
- Vehicle overturning;
- Ride comfort;
- Motion sickness;
- Wear rate on wheel and rail;
- Predict if the wheel profile will be stable or not;
- Predict the risk of out of round wheels.

Special analysis for rail vehicle includes:

- Calculation of roll coefficient;
- Calculation of wheel unloading;
- Calculation of vehicle behaviour in traction and braking.

For example, models of a single passenger car and its bogies are shown in Figure 5.8 [6,7]. Both components of the vehicle body and the bogie frames are modelled as rigid bodies each with 6 DOFs, and the suspension elements used include:

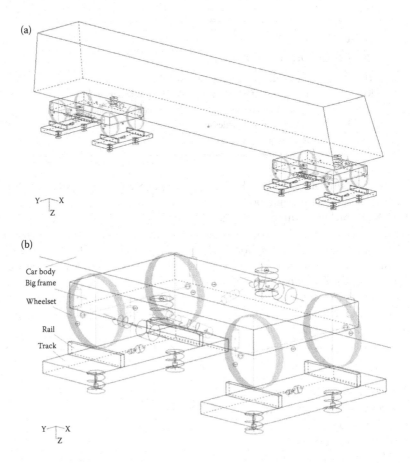

(a)

Y⌐X
⌐Z

(b)

Car body
Big frame

Wheelset

Rail

Track

Y⌐X
⌐Z

FIGURE 5.8 Passenger vehicle model. (a) Full vehicle model, (b) bogie model.

- Two vertical coil spring elements;
- One spring element for the anti-roll bar, and one spring element and one damper with series flexibility for the traction rod in the direction specified by the attachment points of the coupling;
- One lateral and two vertical bumpstops, two vertical viscous dampers, and two lateral viscous dampers and two yaw dampers with series flexibility in the direction specified by the coupling's attachment points, respectively.

5.6.3 VAMPIRE

The owner of VAMPIRE is the DeltaRail Group Ltd, whose predecessors were AEA Technology Rail and British Rail Research, which had been world-leading authorities in the field of rail vehicle dynamics and wheel–rail interaction for many years [8,9]. From that time to now, various methods for the prediction of railway vehicle dynamic behaviour have been developed and carefully validated using sophisticated test facilities and equipment. The various analysis methods have been assembled into a single coherent software package, called VAMPIRE, to allow real problems in railway vehicle dynamics to be solved quickly and cost effectively.

Unlike many other multibody dynamics packages, VAMPIRE is particularly designed to analyse the behaviour of rail vehicles. Therefore, VAMPIRE allows assembling a mathematical model of almost any rail vehicle configuration and offers detailed models of suspension components and elements important to rail vehicle behaviour such as airsprings. It is claimed that running VAMPIRE is significantly faster than other general multibody packages.

VAMPIRE Pro is the latest version of VAMPIRE, and it includes all the pre- and post-processing options, as shown in Table 5.3, required to investigate railway-related issues from vehicle design and acceptance to in-service issues, track damage and accident investigation.

As indicated in Table 5.3, and similar to GENSYS, the VAMPIRE analysis capabilities can be extended through the use of VAMPIRE Control (MATLAB/Simulink® interface) to co-simulate the control algorithms for active or specialist suspensions. More importantly, in order to provide a means for users of VAMPIRE to model and simulate more complex problems, a User Subroutine Facility is available. This facility allows users to write their own algorithms or subroutines to, for example, model and investigate the behaviour of active and other novel suspensions, to simulate control systems, or just to extend the functionality of the standard transient analysis program and so forth.

Figure 5.9 shows a typical three-piece bogie wagon modelling using VAMPIRE [10,11].

The wagon model in Figure 5.9 contains 11 masses (one wagon car body, two bolsters, four sideframes, and four wheelsets). The suspensions among these 11 masses have been modelled using 17 stiffness elements, 74 bumpstop elements, 13 viscous damper elements, 116 friction elements, and four shear spring elements. This model fully considers the nonlinear characteristics of the connections.

TABLE 5.3
Pre- and Post-Processors in VAMPIRE

Pre-processors	Generating and visualising wheel–rail contact data, model building and track plotting
Analysis programs	Linear eigenvalue and response analysis, non-linear transient response analysis, quasi-static curving analysis and static analysis. Can be extended by the use of VAMPIRE user subroutines or VAMPIRE Control (MATLAB/Simulink interface), and the inline processing of simulation data.
Post-processors	Extensive plotting facilities for simulation data, statistical analysis, data filtering, channel arithmetic, data extractor and peak counting. Vehicle acceptance and wheel and rail wear analysis.
Animation	Transient response animations and eigenvalue modes animations.

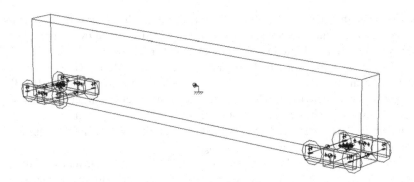

FIGURE 5.9 Wagon model with three-piece bogies in VAMPIRE.

5.6.4 ADAMS/RAIL (NOW VI-RAIL)

MSC ADAMS is one of the leading Multibody Dynamics (MBD) software packages [12,13]. ADAMS/Rail, a specialised simulation software package for railway engineering, is based on the market-leading motion simulation software introduced in 1995. It is currently in use by a number of global railway industry manufacturers. The worldwide railway industry can take full advantage of a virtual prototyping environment featuring the ADAMS/Rail integration platform and a validated real-time tool for conceptual vehicle dynamics.

Railways are seen as a low-cost alternative to air and road transportation. They offer a cost-effective way to move goods and people in large quantities. While the trains may be powered by one of many energy sources, diesel and electricity are the most common sources of energy used. Modernisation of railway systems places increased focus on safety and reliability. Design requirements vary considerably based on the use of the rail vehicle, for example, freight versus passenger, track conditions, maximum speed of the train and so on. ADAMS/Rail offers a complete set of solutions to address the complex problems faced by the railway industry.

ADAMS/Rail software allows rail vehicle engineers to build and test functional virtual prototypes of complex rail vehicle designs, to realistically simulate the full-motion dynamic behaviours on their computers, to evaluate and manage the complex interactions between rail vehicle and tracks, and to better optimise rail vehicle designs for performance, safety and comfort. This helps rail vehicle manufacturers to produce better vehicles faster and at lower cost, with reduced risk and increased communication throughout the vehicle development process. ADAMS/Rail software is relatively easy to use, analysts like the interface and are very comfortable using the system. ADAMS/Rail software can be used for the following types of rail simulations:

- Design optimisation;
- Noise and vibration;
- Acoustics;
- Durability and fatigue;
- Crash and safety;

- Ride and handling;
- Derailment safety clearance;
- Track dynamic loading;
- Traction and braking;
- Structural analysis;
- Thermal performance;
- Mechatronics;
- Vehicle dynamics;
- Multidisciplinary analysis;
- Non-linear analysis;
- Simulation of cargo restraint effectiveness;
- Accident investigation;
- Bridge loading.

Rail vehicle engineers can use ADAMS/Rail software for the following aspects:

- Suspension systems;
- Vehicle frames;
- Power transmission systems;
- Elastomeric seals and mounts;
- Composites modelling and failure analysis;
- Manufacturing processes;
- Energy absorber system design;
- Couplers;
- Control systems;
- Mechatronics;
- Crashworthiness studies;
- Brake systems;
- Cabin noise and comfort;
- Wheel–rail interface.

5.6.5   SIMPACK

As general-purpose software for Multibody Simulation (MBS), SIMPACK can be used for the dynamic analysis of any mechanical or mechatronic system. Its capacity in railway applications enables rail engineers to generate and solve virtual 3D rail vehicle and track models in order to predict and visualise motion, coupling forces and stresses. SIMPACK Rail is particularly well-suited to high-frequency transient analyses, even into the acoustic range [14]. It was primarily developed to handle complex non-linear models with flexible bodies and harsh shock contact. It can be used for the analysis and design of any type of rail-based vehicle and articulated high-speed train. Its applications in railway simulation have:

- Unlimited flexibility;
- Importation of bodies from FEM codes;
- Easy assembly of multi-car trains from submodels;

- Powerful track editor;
- Each wheel treated independently;
- Scalable detail and complexity;
- Batch jobs;
- Automatic analysis reports;
- Curving;
- Rail-to-wheel forces;
- Derailment safety;
- Critical speed;
- Passenger comfort;
- Switch and crossing designs;
- Traction and braking.

5.6.6 UNIVERSAL MECHANISM

Universal Mechanism (UM) is a multibody dynamics program developed at the Laboratory of Computational Mechanics of Bryansk State Technical University, Russia [15]. UM Loco is an additional UM module that is aimed at simulation of the dynamics of rail vehicles, including locomotives, wagons and trains.

UM Loco includes the following configurations in addition to the standard UM ones:

- Wheelset as a standard subsystem;
- Automatic calculation of wheel–rail contact forces according to various models of creep forces (Mueller model, Fastsim, non-Hertzian contact model etc.);
- Specialised graphical interface for animation of contact forces;
- Interface for creating rail and wheel profiles and track irregularities;
- Interface for setting curve and switch parameters;
- Standard list of variables which characterise wheel–rail interaction (creepages, total, normal and creep forces in wheel–rail contacts, angle of attack, wear factors and others, giving more than 30 variables for each vehicle wheel);
- Database of profiles and track irregularities;
- Models of various vehicles.

UM Loco allows the user:

- To calculate the critical speed of a vehicle;
- To analyse 3D dynamics of a vehicle or a train in the time domain on tangent track or in curves with/without irregularities;
- To analyse the vehicle dynamics with regard to dependence on wheel and rail profiles;
- To include 3D vehicles in a train model;
- To create multi-variant projects for scanning the vehicle–train dynamics regarding the dependence on any parameters;

- To compute the natural frequencies and modes, eigenvalue and eigenforms as well as root locus of linearised equations of motion;
- To create hybrid rigid–elastic models of vehicles.

5.6.7 IN-HOUSE PACKAGES

During its early research on train and wagon dynamics, two software packages were developed at the Centre for Railway Engineering (CRE), Central Queensland University around the year 2000. One is called CRE-LTS (CRE—Longitudinal Train Simulator) and the other called CRE-3DWTSD Model (CRE—Three-dimensional Wagon–Track System Dynamics Model).

In CRE-LTS, the train is modelled with non-linear draft gear connections (Figure 5.10). Each vehicle mass is modelled separately connected by the non-linear draft gear connection modelling and slack elements [16,17]. Other forces such as propulsion resistance, curve resistance, grade forces, braking forces and traction forces are added to each vehicle mass as appropriate. Locomotive characteristics of traction and dynamic braking are applied as forces to locomotive vehicle masses and are modelled from manufacturer's performance curves. Air brakes are added as braking forces to all vehicles, but the default braking condition for continuous running is for air brakes to be applied to wagons only. Grade forces can be in either direction and are applied individually to all vehicles. Propulsion resistance is likewise applied to all vehicles and curving resistance added as each vehicle negotiates curves. This gives a system of differential equations which are solved using a numerical Runge–Kutta stepwise integration scheme. As each vehicle is modelled as a separate mass, the multi-degree-of-freedom system has the order equal to the number of vehicles. A more detailed coverage of the modelling can be found in Chapter 6.

In the CRE-3DWTSD model [18] for the lateral and the vertical dynamics of the wagon–track interactions, the wagon subsystem consists of the wagon car body, the bolsters, the secondary suspensions, the sideframes, the primary suspensions and the wheelsets. The total 37 DOFs are used to define the wagon subsystem as shown in Figure 5.11a and b. The track subsystem consists of all track components including the rails, pads, fasteners, sleepers, ballast, subballast and subgrade, and is

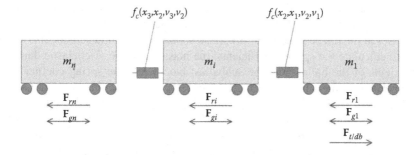

FIGURE 5.10 Longitudinal train modelling.

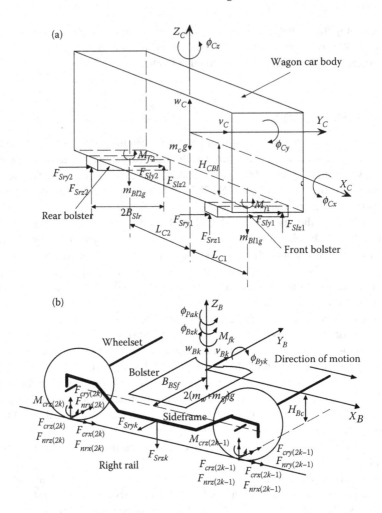

FIGURE 5.11 Wagon model. (a) Wagon body model, (b) bogie model.

modelled as a four-layer track structure (rails, sleepers, ballast blocks and subballast blocks) as shown in Figure 5.12a and b. In the track subsystem, the rail is described using Timoshenko beam theory. The sleeper is considered as a deformable short beam resting on an elastic foundation and represented by its spring and viscoelastic properties at the rail seat location. A ballast–subballast pyramid submodel is developed to calculate the spring coefficients, the masses and the viscoelastic damping coefficients of the ballast and the subballast blocks and the spring coefficient of the subgrade. In the wheel–rail interface system, the normal contact force due to the wheel–rail rolling contact is determined using a modified Hertz static contact theory by including a multi-point contact submodel. The creep forces along the tangential directions and the creep moments about the normal direction are determined using Kalker's linear creep theory. A more detailed coverage of the modelling can be found in Chapter 7.

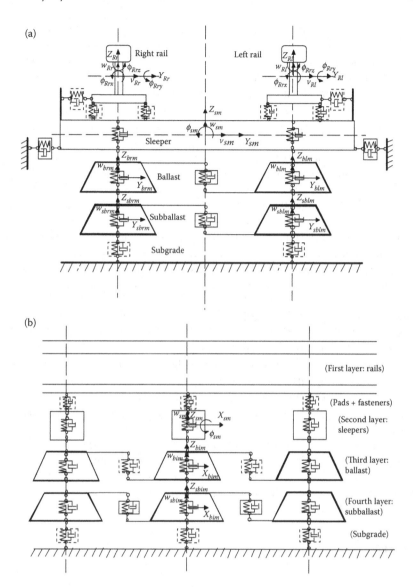

FIGURE 5.12 Track model. (a) Longitudinal view, (b) lateral view.

The dynamic forces that originate from the wheel–rail interface are transmitted up to the wagon body and down to the subgrade. Various track irregularities and defects in the rail and wheel are incorporated in the CRE-3DWTSD model. The predicted vertical and lateral dynamic responses (impacts, hunting and potential for derailment) with and without defects/track irregularities have been validated using the results reported in the literature. The effect of the design and operational parameters of the wagon and track system on the vertical and lateral dynamic responses can be investigated using the CRE-3DWTSD model. From this investigation, overall system dynamic behaviours can be further understood. The predictions of accurate overall system responses from this model can be beneficial to the design of both wagon and track structures. It can be expected that the further investigations of the deterioration of both wagon and track and the optimisation could be completed using this model incorporated with some relevant submodels.

5.6.8 BENCHMARKS AND COMPARISONS

Manchester Benchmarks were developed in 1997 at the International Workshop 'Computer Simulation of Rail Vehicle Dynamics' held in Manchester, UK. The purpose was to let railway vehicle designers and researchers assess the suitability of the various software packages that now exist for simulation of railway vehicle dynamics [19,20].

The two benchmark vehicles, a passenger car and a freight vehicle, were selected and the four track cases were chosen to run with the benchmark vehicle models. Detailed information about the benchmarks, the models, their parameters, track cases and list of evaluated variables can be found in the following book: *The Manchester Benchmarks for Rail Vehicle Simulation*, S. Iwnicki (Editor), Swets & Zeitlinger, 1999, Lisse, The Netherlands, ISBN: 9026515510. The benchmark results for ADAMS/Rail, MEDYNA, GENSYS, NUCARS, SIMPACK and VAMPIRE are also published in this book. The results are presented in the form of tables and plots comparing how each package predicts the vehicle behaviour. These results are discussed and the differences analysed. Among the six packages, three of them (NUCARS, VAMPIRE and GENSYS) showed the best agreement in all cases. The main difference between NUCARS, VAMPIRE and GENSYS was that GENSYS takes the elasticity of the wheel and rail in the contact area into consideration. The elasticity in wheels and rails was an important factor in the benchmarks because the measured profiles caused a non-elliptical contact surface and the actual contact force affected the shape of the contact patch. However, in tests against Manchester Benchmarks with rigid wheels, GENSYS was in very close agreement with VAMPIRE and NUCARS [5].

REFERENCES

1. A.A. Shabana, *Dynamics of Multibody Systems* (3rd ed.), Cambridge University Press, New York, NY, 2005.
2. J.G. De Jalon, E. Bayo, *Kinematic and Dynamic Simulation of Multibody Systems: The Real-Time Challenge*, Springer-Verlag, New York, NY, 1994.
3. S. Iwnicki (Ed.), *Handbook of Railway Vehicle Dynamics*, Taylor & Francis, Boca Raton, FL, 2006.
4. Transportation Technology Center, Inc., NUCARS, 2013. See: http://www.aar.com/nucars/about.asp.
5. AB DEsolver, *GENSYS.1309 Reference Manual*, 2013. See: http://www.gensys.se/ref_man.html.
6. Y.Q. Sun, M. Spiryagin, S. Simson, C. Cole, D. Kreiser, Adequacy of modelling of friction wedge suspension in three-piece bogies, *Proceedings of the 22nd International*

Symposium on Dynamics of Vehicles on Roads and Tracks (CD), Manchester, UK, 14–19 August, 2011.

7. Y.Q. Sun, C. Cole, M. Dhanasekar, Multi-body modelling of wagon train for crashworthiness analysis, *Proceedings of the 22nd International Symposium on Dynamics of Vehicles on Roads and Tracks (CD)*, Manchester, UK, 14–19 August, 2011.

8. L. Rawlings (Ed.), *VAMPIRE (Version 4.32) User Manual*, AEA Technology, Derby, UK, 2004.

9. DeltaRail, *VAMPIRE Pro V6.00 Help Manual*, 2012. See: http://www.vampire-dynamics.com.

10. Y.Q. Sun, C. Cole, P. Boyd, A numerical method using VAMPIRE modelling for prediction of turnout curve wheel–rail wear, *Wear*, 271(1–2), 2011, 482–491.

11. Y.Q. Sun, M. Dhanasekar, D. Roach, Effect of track geometry irregularities on wheel-rail impact forces, *Proceedings of the Conference on Railway Engineering*, Darwin, Australia, 20–23 June, 2004, pp. 03.1–03.7.

12. MSC Software, ADAMS/Rail, 2013. See: http://www.mscsoftware.com/industry/rail.

13. G. Ferrarotti, ADAMS/Rail 10.1: A revolutionary environment for railway vehicle simulation, *5th Adams Rail Users Conference*, Haarlem, The Netherlands, 10–12 May, 2000.

14. INTEC GmbH, *SIMPACK Rail, Simpack Release 8.9*, 2008. See: http://www.simpack.com/fileadmin/simpack/doc/newsletter/2008/sn-1-08-SIMPACK_8.9_new_functionality.pdf.

15. Universal Mechanism, *User's Manual—Simulation of Rail Vehicle Dynamics—Version 7.0*, 2013, Laboratory of Computational Mechanics, Bryansk State Technical University, Russia. See: http://www.universalmechanism.com/en/pages/index.php?id=3.

16. C. Cole, Longitudinal train dynamics: Characteristics, modelling, simulation and neural network prediction for Central Queensland coal trains, PhD thesis, Central Queensland University, Rockhampton, Australia, 1999. See: http://hdl.cqu.edu.au/10018/928084.

17. Y.Q. Sun, A wagon–track system dynamics model for the simulation of heavy haul railway transportation, PhD thesis, Central Queensland University, Rockhampton, Australia, 2003. See: http://trove.nla.gov.au/version/27636656.

18. C. Cole, Y.Q. Sun, Simulated comparisons of wagon coupler systems in heavy haul trains, *Rail and Rapid Transit*, 220(3), 2006, 247–256.

19. S. Iwnicki, Manchester benchmarks for rail vehicle simulation, *Vehicle System Dynamics*, 30(3–4), 1998, 295–313.

20. P. Shackleton, S. Iwnicki, Comparison of wheel–rail contact codes for railway vehicle simulation: An introduction to the Manchester Contact Benchmark and initial results, *Vehicle System Dynamics*, 46(1–2), 2008, 129–149.

6 Longitudinal Train Dynamics

6.1 INTRODUCTION TO LONGITUDINAL TRAIN DYNAMICS

This chapter has been designed to provide a hands-on guide to both understanding and analysing longitudinal train dynamics. The chapter is intentionally more general than a previous chapter by the author [1], which concentrates mainly on freight trains, non-linear friction-type draft gears and heavy haul issues. For more details on these issues, the reader is referred to [1]. This chapter provides a step-by-step journey through the modelling process. To provide a more instructive resource, it was decided to add modelling using the popularly available Simulink® package. In this way, those wishing to understand train dynamics more deeply and those who wish to use the book as an instruction resource can work through these models progressively.

Longitudinal train dynamics is defined as the motions of rolling stock vehicles in the direction along the track. It therefore includes the motion of the train as a whole and any relative motions between vehicles allowed due to the looseness and travel allowed by spring and damper connections between vehicles. In the railway industry, the relative motions of vehicles is known as 'slack action' due to the correct understanding that these motions are primarily allowed by the free slack and deflections allowed in wagon connections. Coupling 'free slack' is defined as the free movement allowed by the sum of the clearances in the wagon connection. In the case of auto-couplers, these clearances consist of clearances in the auto-coupler knuckles and draft gear assembly pins. In older rolling stock connection systems, such as drawhooks and buffers, free slack is the clearance between the buffers measured in tension. Note that a system with drawhooks and buffers could be preloaded with the screw link to remove free slack. The occurrence of 'slack action' is further classified in various railways by various terms; in the Australian industry vernacular, the events are referred to as 'run-ins' and 'run-outs'. The case of a 'run-in' describes the situation where vehicles are progressively impacting each other as the train compresses. The case of a 'run-out' describes the opposite situation where vehicles are reaching the extended extreme of connection-free slack as the train stretches. In other countries different terms are used, for example, impact accelerations, jerk and so forth. Longitudinal train dynamics therefore has implications for passenger comfort, vehicle stability, rolling stock design and rolling stock metal fatigue [1].

The study and understanding of longitudinal train dynamics was probably firstly motivated by the desire to reduce the longitudinal vehicle dynamics in passenger trains and, in so doing, improve the general comfort of passengers. The practice of 'power braking', which is the seemingly strange technique of keeping the locomotive power applied while a minimum air brake application is made, is still practised widely on passenger trains. Power braking is also used on partly loaded mixed freight trains to keep the train stretched during braking and when operating on undulating track. Interest in train dynamics in freight trains increased as trains became longer, particularly for heavy haul trains as evidenced in technical papers. In the late 1980s, measurement and simulation of in-train forces on such trains was reported by Duncan and Webb [2]. The engineering issues associated with moving to trains of double the existing length was reported at the same time in New South Wales in a paper by Jolly and Sismey [3]. A further paper focused on train handling techniques on the Richards Bay Line gave the South Africa experience [4]. The research at this time was driven primarily by the occurrences of fatigue cracking and tensile failures in auto-couplers. From these studies, an understanding of the force magnitudes was developed along with an awareness of the need to limit these forces with appropriate driving strategies [1–4].

More recent research into longitudinal train dynamics was started in the early 1990s motivated, not this time by equipment failures and fatigue damage, but by derailments. The direction of this research was concerned with the linkage of longitudinal train dynamics to increases in wheel unloading. It stands to reason that, as trains get longer and heavier, in-train forces get larger. As coupler forces became larger, so too did the lateral and vertical components of these forces resulting from coupler angles on horizontal and vertical curves. At some point, these components will adversely affect wagon stability. The first known work published addressing this issue was that of El-Sibaie [5], which looked at the relationship between lateral coupler force components and wheel unloading. Further modes of interaction were reported and simulated by McClanachan et al. [6] detailing wagon body and bogie pitch.

Concurrent with this emphasis on the relationship between longitudinal dynamics and wagon stability is the emphasis on train energy management. The operation of larger trains meant that the energy consequences for stopping a train became more significant. Train simulators were also applied to the task of training drivers to reduce energy consumption. Measurements and simulations of energy consumed by trains normalised per kilometre-tonne hauled have shown that different driving techniques can cause large variances in the energy consumed [7,8].

6.2   MODELLING LONGITUDINAL TRAIN DYNAMICS

6.2.1   TRAIN MODELS

The longitudinal behaviour of trains is a function of train control inputs from the locomotive, train brake inputs, track topography, track curvature, rolling stock and bogie characteristics and wagon connection characteristics.

The longitudinal dynamic behaviour of a train can be described by a system of differential equations. For the purposes of setting up the equations for modelling and simulation, it is usually assumed that there is no lateral or vertical movement of the wagons. This simplification of the system is employed by all known rail specific, commercial simulation packages and by texts such as Dukkipati and Garg [9]. The governing differential equations can be developed by considering the generalised

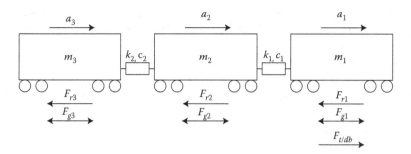

FIGURE 6.1 Three-mass train model.

three mass trains in Figure 6.1. It will be noticed that the in-train vehicle, whether locomotive or wagon, can be classified as one of only three connection configurations, lead (shown as m_1), in-train and tail. All vehicles are subject to retardation and grade forces. Traction and dynamic brake forces are added to powered vehicles.

In Figure 6.1, a = vehicle acceleration, m/s^2; c = damping constant, N · s/m; k = spring constant, N/m; m = vehicle mass, kg; v = vehicle velocity, m/s; x = vehicle displacement, m; F_g = gravity force components due to track grade, N; F_r = sum of retardation forces, N; and $F_{t/db}$ = traction and dynamic brake forces from a locomotive unit, N.

It will be noted on the model in Figure 6.1 that the grade force can be in either direction. The sum of the retardation forces, F_r, is made up of rolling resistance, curving resistance or curve drag, air resistance and braking (excluding dynamic braking which is more conveniently grouped with locomotive traction in the $F_{t/db}$ term). Rolling and air resistances are usually grouped as a term known as propulsion resistance, F_{pr}, making the equation for Fr as follows:

$$F_r = F_{pr} + F_{cr} + F_b$$

where F_{pr} is the propulsion resistance, F_{cr} is the curving resistance and F_b is the braking resistance due to pneumatic braking.

The three mass train allows the three different differential equations to be developed. With linear wagon connection models, the equations can be written as

$$m_1 a_1 + c_1(v_1 - v_2) + k_1(x_1 - x_2) = F_{t/db} - F_{r1} - F_{g1} \qquad (6.1)$$

$$m_2 a_2 + c_1(v_2 - v_1) + c_2(v_2 - v_3) + k_1(x_2 - x_1) + k_2(x_2 - x_3) = F_{r2} - F_{g2} \qquad (6.2)$$

$$m_3 a_3 + c_2(v_3 - v_2) + k_2(x_3 - x_2) = -F_{r3} - F_{g3} \qquad (6.3)$$

Note that a positive value of F_g is taken as an upward grade, that is, a retarding force.

Allowing for locomotives to be placed at any train position and extending equation notation for a train of any number of vehicles, a more general set of equations can be written as follows:

For the lead vehicle:

$$m_1 a_1 + c_1(v_1 - v_2) + k_1(x_1 - x_2) = F_{t/db1} - F_{r1} - F_{g1} \qquad (6.4)$$

For the ith vehicle:

$$m_i a_i + c_{i-1}(v_i - v_{i-1}) + c_i(v_i - v_{i+1}) + k_{i-1}(x_i - x_{i-1}) + k_i(x_i - x_{i+1}) = \\ F_{t/dbi} - F_{ri} - F_{gi} \qquad (6.5)$$

For the nth or last vehicle:

$$m_n a_n + c_{n-1}(v_n - v_{n-1}) + k_{n-1}(x_n - x_{n-1}) = F_{t/dbn} - F_{rn} - F_{gn} \qquad (6.6)$$

By including the $F_{t/db}$ in each equation, and thus on every vehicle, the equations can be applied to any locomotive placement or system of distributed power. For unpowered vehicles, $F_{t/db}$ is set to zero.

For non-linear modelling of the system, the stiffness and damping constants are replaced with functions or more complex non-linear models. In the general case, the model must include dependency on both displacement and velocity (see Figure 6.2). The generalised non-linear equations are therefore:

For the lead vehicle:

$$m_1 a_1 + f_{wc}(v_1, v_2, x_1, x_2) = F_{t/db1} - F_{r1} - F_{g1} \qquad (6.7)$$

For the ith vehicle:

$$m_i a_i + f_{wc}(v_i, v_{i-1}, x_i, x_{i-1}) + f_{wc}(v_i, v_{i+1}, x_i, x_{i+1}) = F_{t/dbi} - F_{ri} - F_{gi} \qquad (6.8)$$

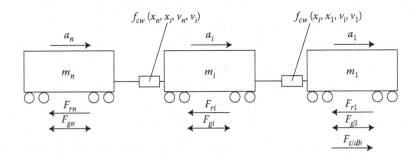

FIGURE 6.2   Generalised train model.

For the nth or last vehicle:

$$m_n a_n + f_{wc}(v_n, v_{n-1}, x_n, x_{n-1}) = F_{t/dbn} - F_{rn} - F_{gn} \qquad (6.9)$$

where f_{wc} is the non-linear function describing the full characteristics of the wagon connection.

Solution and simulation of the above equation set is further complicated by the need to calculate the forcing inputs to the system, that is, $F_{t/db}$, F_r and F_g. The traction-dynamic brake force term, $F_{t/db}$, must be continually updated for driver control adjustments and any changes to locomotive speed. The retardation forces, F_r, are dependent on braking settings, velocity, curvature and rolling stock design. Gravity force components, F_g, are dependent on track grade and therefore the position of the vehicle on the track. Approaches to the non-linear modelling of the wagon connection and modelling of each of the forcing inputs are included and discussed in the following sections.

6.2.2 Modelling Vehicle Inputs

As it is intended to give a hands-on guide, it is logical to develop all the modelling that could be associated with each individual vehicle before assembling the whole train model. The single vehicle model is firstly developed as described by the equation:

$$m_1 a_1 = F_{t/db} - F_{r1} - F_{g1} \qquad (6.10)$$

Note that this is just Equation 6.1 with the wagon connection removed. A simple Simulink model is given in Figure 6.3.

The single vehicle model has a mass of 120 tonnes and is provided with a constant traction input of 10 kN. The only resistive force in this case is rolling and air

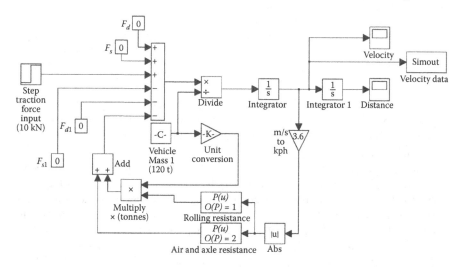

FIGURE 6.3 Single vehicle model—implemented in Simulink.

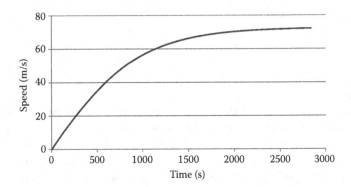

FIGURE 6.4 Simulation results—speed response of single vehicle model.

resistance. This rather hypothetical case reaches 'terminal' velocity after approximately 2500 s, as shown in Figure 6.4.

6.2.2.1 Locomotive Traction and Dynamic Braking

Of course, the modelling in Figure 6.3 is oversimplified. Traction, braking and other input forces are not provided as step inputs. Locomotive traction and dynamic braking have evolved over many years and several systems exist. In diesel locomotives, a tradition of eight notches for the throttle control emerged based on a three-valve fuel control. More modern locomotives can have different numbers of notches and levels for dynamics braking, although eight remains common for operational reasons. As designs have become complex, it is now usual to base models upon manufacturers' locomotive performance curves. An approximate model for traction, assuming that notch level is linearly proportional to motor current, can be derived from the following equations:

$$\text{For} \quad F_{t/db} * v < (N^2/64) * P_{max}, \quad F_{t/db} = (N/8) * Te_{max} - k_f * v \qquad (6.11)$$

$$\text{Else,} \quad F_{t/db} = (N^2/64) * P_{max}/v \qquad (6.12)$$

where N is the throttle setting in notches, 0–8; P_{max} is the maximum locomotive traction horsepower, W; Te_{max} is the maximum locomotive traction force, N and k_f is the torque reduction, N/(m/s). Equations 6.11 and 6.12 adequately describe locomotives as shown in Figure 6.5.

While a reasonable fit to the published power curves may be possible with a simple equation of the form $P = F_{t/db} * v$, it may be necessary to modify this model to reflect further control features or reflect the changes in efficiency or thermal effects at different train speeds. It is common for the traction performance characteristic to fall below the power curve $P_{max} = F_{t/db} * v$ at higher speeds due to limits imposed by the generator maximum voltage. Enhanced performance closer to the power curve at higher speeds is achieved on some locomotives by adding a motor field weakening control [10]. It can be seen that accurate modelling of locomotives, even without

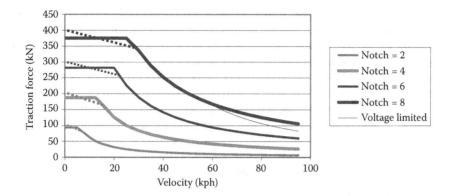

FIGURE 6.5 Typical tractive effort performance curves—diesel electric. (From C. Cole, in: *Handbook of Railway Vehicle Dynamics*, Chapter 9, Taylor & Francis, Boca Raton, FL, pp. 239–278, 2006. With permission.)

considering the electrical modelling in detail, can become quite complicated. In all cases, the performance curves should be sourced and as much precise detail as possible should be obtained about the control features to ensure a suitable model is developed.

It is typical for locomotive manufacturers to publish both the maximum tractive effort and the maximum continuous tractive effort. The maximum continuous tractive effort is the traction force delivered at full throttle notch after the traction system has heated to a nominal maximum operating temperature. As the resistivity of the windings increases with temperature, the motor torque decreases (i.e. due to the lower motor current). As traction motors have considerable mass, considerable time is needed for the locomotive motors to heat with performance levels progressively dropping to maximum continuous tractive effort. A typical thermal derating curve for a modern locomotive is shown in Figure 6.6.

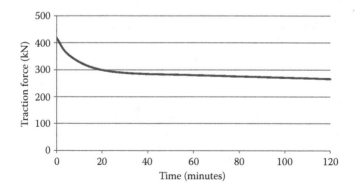

FIGURE 6.6 Tractive effort thermal derating curve. (From C. Cole, in: *Handbook of Railway Vehicle Dynamics*, Chapter 9, Taylor & Francis, Boca Raton, FL, pp. 239–278, 2006. With permission.)

Manufacturers' data from which performance curves such as those shown in Figure 6.5 are derived can usually be taken to be maximum rather than continuous values. If the longitudinal dynamics problem under study has severe grades and locomotives are delivering large traction forces for long periods, it will be necessary to modify the simple model represented in Figure 6.5 with a further model adding these thermal effects.

A key parameter in any discussion about tractive effort is rail–wheel adhesion or the coefficient of friction. Prior to enhancement of motor torque control, a rail–wheel adhesion level of ~0.20 could be expected. Modern locomotive traction control systems deliver higher values of adhesion reaching ~0.35 in daily operation with manufacturers claiming up to 0.52 in published performance curves. It needs to be remembered that a smooth control system can only deliver an adhesion level up to the maximum set by the coefficient of friction for the wheel–rail conditions. Wheel–rail conditions in frost and snow could reduce adhesion to as low as 0.1. Superimposing adhesion levels in Figure 6.5, as shown in Figure 6.7, indicates how significant adhesion is as a locomotive performance parameter.

The use of dynamic brake as a means of train deceleration has continued to increase as dynamic brake control systems have been improved. Early systems as shown in Figure 6.8 gave only a variable retardation force and were not well received by drivers. As the effectiveness was so dependent on velocity, the use of dynamic brake gave unpredictable results unless a mental note was made of locomotive velocity and the driver was aware of what performance to expect. Extended range systems, which involved switching resistor banks, greatly improved dynamic brake usability on diesel electric locomotives. More recent locomotive packages have provided large regions of maximum retardation at steady force levels. The performance of the dynamic brake is limited at higher speeds by current, voltage and commutator limits. Performance at low speeds is limited by the motor field. Designs have continued to extend the full dynamic brake force capability to lower and lower speeds. Recent designs have achieved the retention of maximum dynamic braking force down to

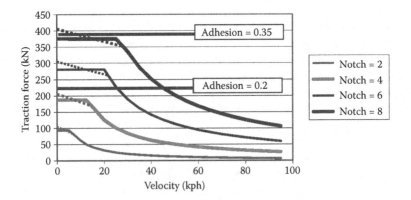

FIGURE 6.7 Tractive effort performance curves—showing effect of adhesion levels. (From C. Cole, in: *Handbook of Railway Vehicle Dynamics*, Chapter 9, Taylor & Francis, Boca Raton, FL, pp. 239–278, 2006. With permission.)

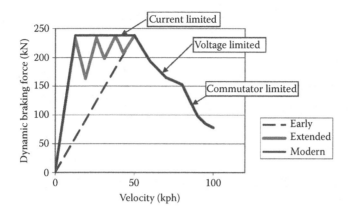

FIGURE 6.8 Dynamic brake characteristics. (From C. Cole, in: *Handbook of Railway Vehicle Dynamics*, Chapter 9, Taylor & Francis, Boca Raton, FL, pp. 239–278, 2006. With permission.)

2 kph. Dynamic brake can be controlled as a continuous level or at discrete control levels, depending on the locomotive. The way in which the control level affects the braking effort differs for different locomotive traction packages. Four different dynamic brake characteristics have been identified, but further variations are not excluded, see Figures 6.9 and 6.10.

The later designs (shown on the left in Figures 6.9 and 6.10) provide larger ranges of speed where a near constant braking effort can be applied. Modelling of the characteristic can be achieved by fitting a piecewise linear function to the curve representing 100% dynamic braking force. The force applied to the simulation can then be scaled linearly in proportion to the control setting. In some configurations, it will be necessary to truncate the calculated value by different amounts. Some examples of dynamic brake characteristics are given in Figures 6.9 and 6.10.

While traction characteristics can sometimes be reduced to a small number of equations, dynamic braking is usually more complicated, requiring piecewise functions and/or a lookup table. A simplified locomotive model is developed in Simulink

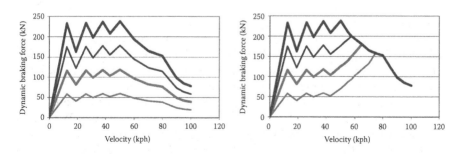

FIGURE 6.9 Dynamic brake characteristics—diesel electric locomotives. (From C. Cole, in: *Handbook of Railway Vehicle Dynamics*, Chapter 9, Taylor & Francis, Boca Raton, FL, pp. 239–278, 2006. With permission.)

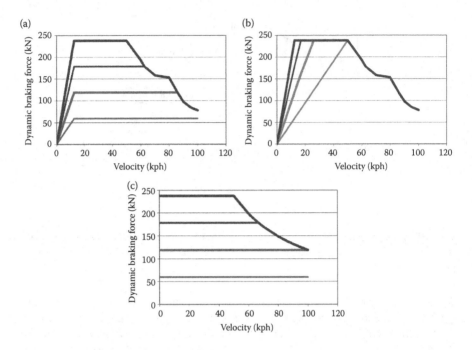

FIGURE 6.10 Dynamic brake characteristics—electric (From C. Cole, in: *Handbook of Railway Vehicle Dynamics*, Chapter 9, Taylor & Francis, Boca Raton, FL, pp. 239–278, 2006.) and AC diesel electric locomotives. (a) Modern DC electric locomotive, (b) older-type DC electric locomotive, (c) modern AC diesel–electric or electric locomotive.

in Figure 6.11 with its outputs shown in Figure 6.12. Again, oversimplified control inputs are added to demonstrate the model.

To continue the process of building a train model, the locomotive traction model of Figure 6.11 is added to the single vehicle model to give a locomotive model as shown in Figure 6.13, with model output results shown in Figure 6.14.

6.2.2.2   Propulsion Resistance

Propulsion resistance has already been introduced indirectly at the beginning of this section in the model in Figure 6.3 and results in Figure 6.4.

Propulsion resistance is usually defined as the sum of rolling resistance and air resistance. In most cases, increased vehicle drag due to the track curvature is considered separately. The variable shapes and designs of rolling stock and the complexity of aerodynamic drag mean that the calculation of rolling resistance is still dependent on empirical formulas. Typically, propulsion resistance is expressed in an equation of the form $R = A + BV + CV^2$. Hay presents the work of W.J. Davis which identifies the term A as journal resistance dependent on both wagon mass and the number of axles; an equation of the form $R = ax + b$, giving in imperial units $1.3wn + 29n$ where w is the weight per axle and n is the number of axles, is quoted in [11]. The second term is mainly dependent on flanging friction and the

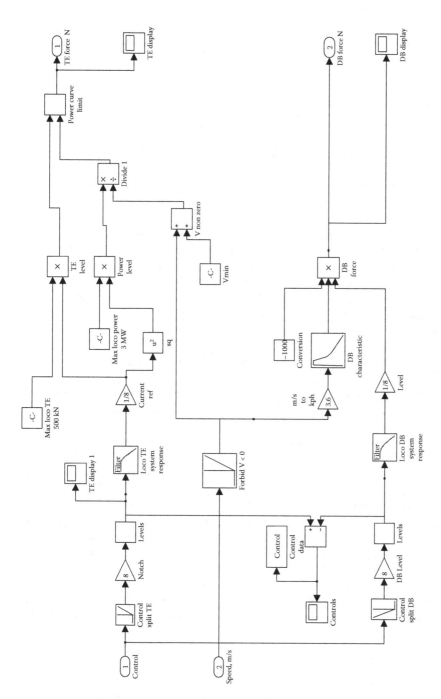

FIGURE 6.11 Locomotive traction dynamic brake model—implemented in Simulink.

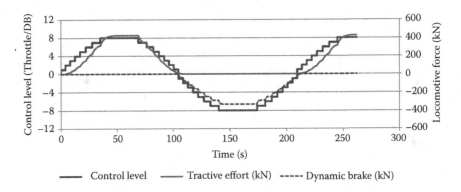

FIGURE 6.12   Simulation results—locomotive traction dynamic brake at speed 7 m/s.

coefficient B is therefore usually small (non-existent in some empirical formulas) and the third term is dependent on air resistance. The forms of propulsion resistance equations used and the empirical factors selected vary between railway systems, reflecting the use of equations that more closely match the different types of rolling stock and running speeds. There are many variations of propulsion resistance equations as shown by Equations 6.13 and 6.14. An instructive collection of propulsion resistance formulas has been assembled in Tables 6.1 and 6.2 from [11] and work by Profillidis [12]. All equations are converted into SI units and expressed as Newtons per tonne mass. A graphical representation of their various outcomes is provided in Figure 6.15.

In Tables 6.1 and 6.2, K_a is an adjustment factor depending on rolling stock type; k_{ad} is an air drag constant depending on car type; m_a is mass supported per axle in tonnes; n is the number of axles; V is the velocity in kilometres per hour and ΔV is the headwind speed, usually taken as 15 kph.

Even with the number of factors described in these tables, the effects of many factors are not and usually cannot be meaningfully considered. How are the differing rolling stock frontal and side areas considered? How are headwind, crosswind and tailwind considered? How are the drag forces from poor bogie steering considered? In the area of air resistance, wagon body design is more variable than suggested by the few adjustment factors presented here. Higher than expected aerodynamic drag has been observed from the addition of headwinds with a slight crosswind component for certain types of trains (e.g. open empty hopper wagons). The dynamicist should therefore be aware that considerable difference between calculations and field measurements is probable. Similarly, in regard to bogie steering and drag, the equations do not include centre bowl friction, warp stiffness or wheel and rail profile information.

6.2.2.3   Curving Resistance

Curving resistance calculations have similarity to propulsion resistance calculations in that empirical formulas must be used. Rolling stock design and condition, cant (superelevation) deficiency, rail profile, rail lubrication and curve radius will all

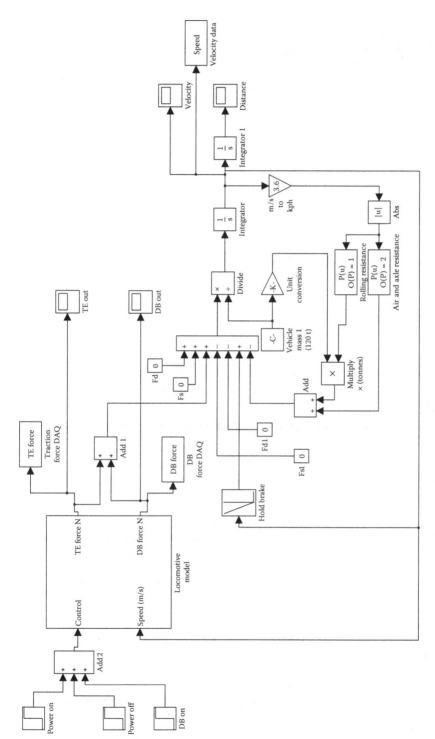

FIGURE 6.13 Locomotive traction dynamic brake vehicle model—implemented in Simulink.

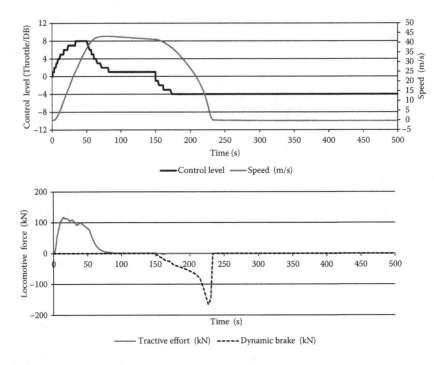

FIGURE 6.14 Simulation results—locomotive traction dynamic brake vehicle.

TABLE 6.1
Empirical Formulas for Propulsion Resistance—Freight Rolling Stock

Description	Equation 6.13
Original Davis equation (USA)	$6.378 + 124.6/m_a + BV + CAV^2/(m_a n)$ A = frontal area in square metres $B = 0.03$ locomotives, 0.045 freight cars $C = 0.0024$ locomotives, 0.0005 freight cars
Modified Davis equation (USA)	$K_a[2.943 + 89/m_a + 0.0305V + 1.718k_{ad}V^2/(m_a\,n)]$ $K_a = 1.0$ for pre-1950, 0.85 for post-1950, 0.95 container on flat car, 1.05 trailer on flat car, 1.05 hopper cars, 1.2 empty covered auto racks, 1.3 for loaded covered auto racks and 1.9 for empty, uncovered auto racks. $k_{ad} = 0.07$ for conventional equipment, 0.0935 for containers and 0.16 for trailers on flat cars
French locomotives	$0.65m_a n + 13n + 0.01m_a nV + 0.03V^2$
French standard UIC vehicles	$9.81(1.25 + V^2/6300)$
French express freight	$9.81(1.5 + V^2/(2000..\,2400))$
French 10t/axle	$9.81(1.5 + V^2/1600)$
French 18t/axle	$9.81(1.2 + V^2/4000)$
German Strahl formula	$25 + k(V + \Delta V)/10$ $k = 0.05$ for mixed freight trains, 0.025 for block trains
Broad gauge (i.e. 1.676 m)	$9.81[0.87 + 0.0103V + 0.000056V^2]$
Broad gauge (i.e. ~1.0 m)	$9.81[2.6 + 0.0003V^2]$

TABLE 6.2
Empirical Formulas for Propulsion Resistance: Passenger Rolling Stock

Description	Equation 6.14
French passenger on bogies	$9.81(1.5 + V^2/4500)$
French passenger on axles	$9.81(1.5 + V^2/(2000.. 2400))$
French TGV	$2500 + 33\,V + 0.543\,V^2$
German Sauthoff formula freight (Intercity Express, ICE)	$9.81[1 + 0.0025V + 0.0055 * ((V + \Delta V)/10)^2]$
Broad gauge (i.e. 1.676 m)	$9.81[0.6855 + 0.02112V + 0.000082V^2]$
Narrow gauge (i.e. ~1.0 m)	$9.81[1.56 + 0.0075V + 0.0003V^2]$

affect the resistance imposed on a vehicle on the curve. As rolling stock design and condition can vary, as can the cant deficiency and rail profile, it is usual to estimate the curving resistance by a function relating only to curve radius. The equation commonly used, as detailed in [11], is

$$F_{cr} = 6116/R \qquad (6.15)$$

where F_{cr} is in Newtons per tonne of wagon mass and R is the curve radius in metres.

Rail flange lubrication is thought to be capable of reducing curving resistance by 50%. The curving resistance of a wagon that is stationary on a curve is thought to be approximately double the value given by Equation 6.15.

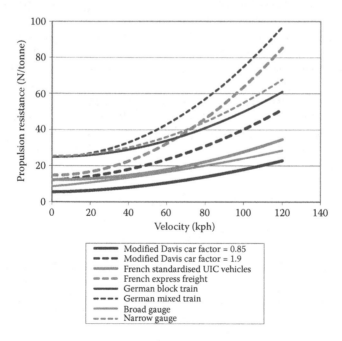

FIGURE 6.15 Propulsion resistance equations compared—freight rolling stock.

6.2.2.4 Gravitational Components

Gravitational components, F_g, are added to longitudinal train models by simply resolving the weight vector into components parallel and at right angles to the wagon body chassis. The parallel component of the vehicle weight becomes F_g. On a grade, a force will either be added to or subtracted from the longitudinal forces on the wagon (see Figures 6.1 and 6.16).

The grade also reduces the sum of the reactions of the wagon downward on the track. This effect has implications for propulsion resistance equations that are dependent on vehicle weight. The effect is, however, small and due to the inherent uncertainty in propulsions resistance calculations it can be safety ignored. Taking a 1 in 50 grade as an example gives a grade angle of 1.146°. The cosine of this angle is 0.99979. The reduction in the sum of the normal reactions for a wagon on a 1 in 50 grade is therefore a fraction of 0.0002 or 0.02%. Grades are obtained from track plan and section data. The grade force component must be calculated for each vehicle in the train and updated each time step during simulation to account for train progression along the track section.

Track grades are added to the Simulink model in Figure 6.17 with results in Figure 6.18. Curve drag is added in Figure 6.19 with results in Figure 6.20. It will be noted that the curve drag of the few curves had very little effect on the results.

6.2.2.5 Pneumatic Brake Models

The modelling of the brake system requires the simulation of a fluid dynamic system that must run in parallel with the train simulation. The output from the brake pipe simulation is the brake cylinder force, which is converted by means of rigging factors and shoe friction coefficients into a retardation force that is one term of the sum of retardation forces F_r.

Modelling of the brake pipe and triple valve systems is a subject in itself, and will therefore not be treated in this chapter beyond characterising the forces that can be expected and the effect of these forces on train dynamics. The majority of freight rolling stock still utilises pneumatic control of the brake system. The North American system differs in design from the Australian/British systems, but both apply brakes sequentially starting from the points at which the brake pipe is exhausted. Both systems depend on the fail-safe feature whereby the opening of the brake valve in the

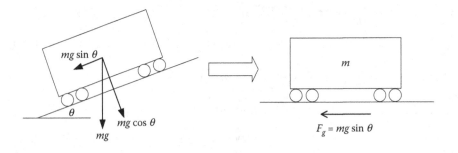

FIGURE 6.16 Modelling gravitational components.

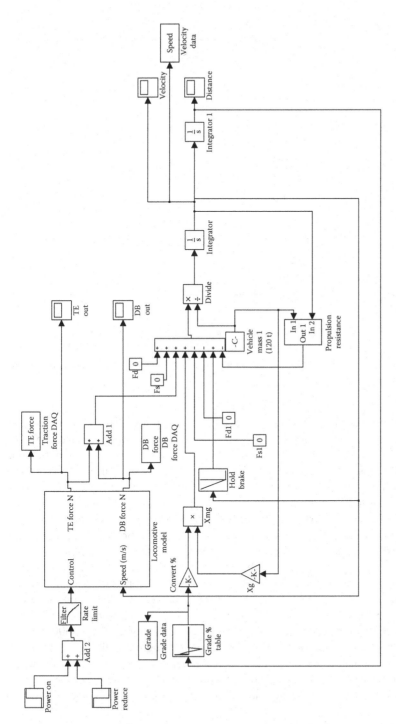

FIGURE 6.17 Locomotive traction dynamic brake vehicle model with grades—implemented in Simulink.

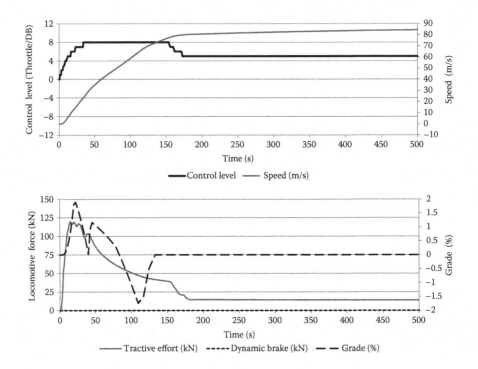

FIGURE 6.18   Simulation results—locomotive traction dynamic brake vehicle model with grades.

locomotive or the fracture of the brake pipe allowing loss of brake pipe pressure results in application of brakes in the train. The particular valves used on each wagon to apply the brakes all work on the same principle but will vary slightly in function and capabilities. The Australian/British systems tend to name these valves 'Triple valves', while they are known as 'AB valves' in North America and 'Distributor valves' in Europe.

Irrespective of the particular version of pneumatically controlled brakes, the key issue is that the pneumatic control adjustments made to the brakes via the brake pipe take time to propagate along the train. Since the first triple valve systems were introduced in the late 1800s, many refinements have been progressively added to ensure or improve brake control propagation. As the control is via a pressure wave, the system is limited to sonic speed which is 350 m/s for sound in air (noting 318 m/s at −20°C, 349 m/s at 30°C). Allowing for losses in brake equipment, a well-designed system can achieve signal propagation at speeds typically in the range of 250–300 m/s. For short trains of 20 wagons (each 15 m long, ~300 m long train), this gives quite reasonable performance. As trains have increased in length, in particular for heavy haul applications (lengths of 1.6–4.0 km), brake control signal propagation can take several seconds. Some simulated data of a brake system emergency application is given in Figure 6.21.

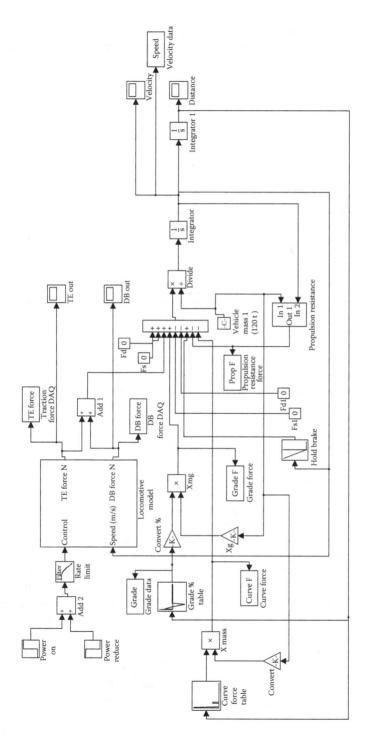

FIGURE 6.19 Locomotive traction dynamic brake vehicle model with grades and curves—implemented in Simulink.

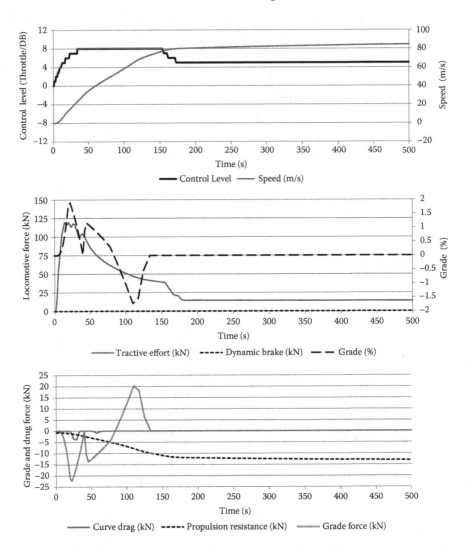

FIGURE 6.20 Simulation results—locomotive traction dynamic brake vehicle model with grades and curves.

It is the demand for better braking in these longer trains that is the primary driver for recent adoption of electronically controlled brakes, which can apply all train brakes almost simultaneously. To introduce readers to the issues of brake control, a Simulink model of the function of a triple valve has been introduced in Figure 6.22. The very simplified model presented assumes air flow through an orifice (in practice, much more precise controls exist using the complicated brake triple valve mentioned earlier, with flows precisely controlled with an arrangement of orifices and chokes), giving:

$$v = K \, (\Delta P)^{0.5} \qquad\qquad (6.16)$$

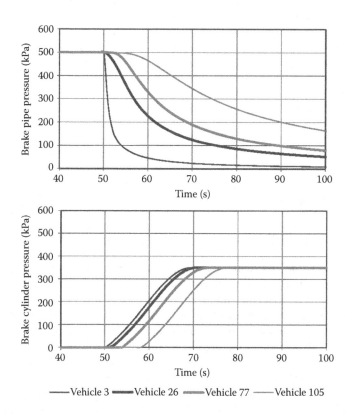

FIGURE 6.21 Simulation results—brake pipe and cylinder responses—emergency application. (From C. Cole, in: *Handbook of Railway Vehicle Dynamics*, Chapter 9, Taylor & Francis, Boca Raton, FL, pp. 239–278, 2006.)

where v is flow velocity, K is the proportionality constant, ΔP is the difference between reservoir pressure and cylinder pressure. Pressure on the cylinder can be estimated from Boyle's law, $PV = mRT$, where P is pressure, V is volume, m is mass, R is the gas constant for air and T is the absolute temperature. As V is the cylinder volume which can be taken as constant after initial movement, and T is the assumed constant, the equation can be written as

$$P = mRT/V; \text{ or differential form } dP = dmRT/V \qquad (6.17)$$

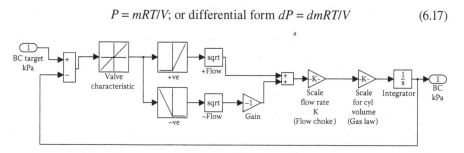

FIGURE 6.22 Simplified triple valve cylinder model—implemented in Simulink.

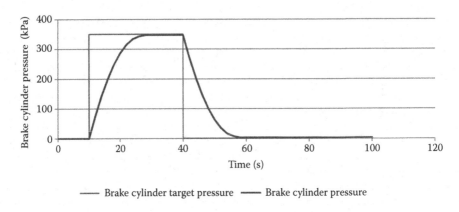

— Brake cylinder target pressure — Brake cylinder pressure

FIGURE 6.23   Simulation results—simplified triple valve cylinder model.

By integrating the differential form of Equation 6.17 and using the gas flow rate from Equation 6.16, the pressure rise in the cylinder can be modelled. Results from this very simplified model are shown in Figure 6.23. While it is acknowledged that this model is greatly simplified and many assumptions are quite 'rough', the characteristic curve shown in Figure 6.23 is quite representative. Responses can be easily tuned to match the dynamics required.

This simplified model is then implemented in a three-vehicle train brake pipe model as shown in Figure 6.24. The propagation delays in the pipe signal, along with the differences in brake pipe drop rates, are modelled using standard delays and low-pass filters from the Simulink library. The simulation results are shown in Figures 6.25 and 6.26. Again, these are not exact models of the actual pipe characteristic, but are useful representations for the purpose of study. The low-pass filters provide a quick and easy way to model brake pipe response time; again, in practice, more exact models can be developed.

Note that, for a three vehicle model, delays are minimal and the delays shown in Figure 6.25 are exaggerated. To better illustrate the delay issue, the model is simulated again assuming that the third vehicle is 750 m away; results are shown in Figure 6.26.

As shown in Figure 6.26, the cylinder fill rates for the brake cylinder at the tail of the train are now limited by the control target provided by the brake pipe rather than the fill rates allowed by the chokes in the triple valve systems. This problem tends to limit the maximum length of brake pipe systems and is a reason for the interest in electronically controlled braking for long heavy haul trains. An electronically controlled brake system is modelled in Figure 6.27, with results in Figure 6.28. Control delays are removed, but note that brakes still have a fill time as shown in Figure 6.28.

6.2.3   WAGON CONNECTION MODELS

Perhaps the most important component in any longitudinal train simulation is the wagon connection element. The auto-coupler with friction type draft gears is the most common wagon connection in the Australian and North American freight train

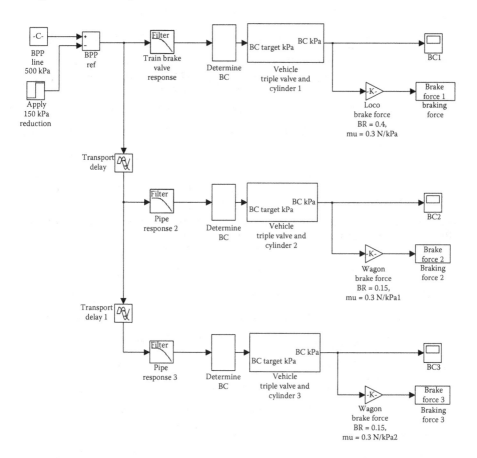

FIGURE 6.24 Simplified three-vehicle train air brake model—implemented in Simulink.

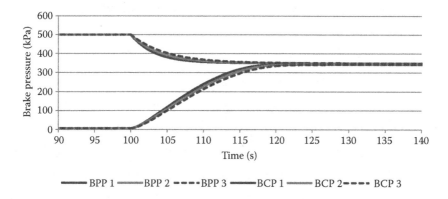

FIGURE 6.25 Simulation results—simplified three-vehicle train air brake model. (BPP = brake pipe pressure, BCP = brake cylinder pressure).

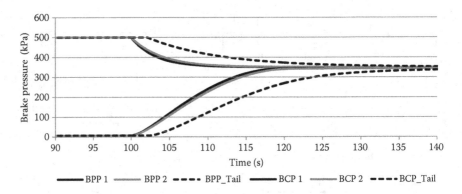

FIGURE 6.26 Simulation results—simplified train air brake model for 750 m brake pipe. (BPP = brake pipe pressure, BCP = brake cylinder pressure).

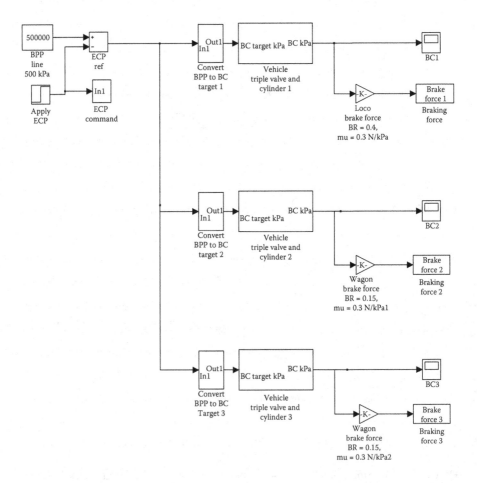

FIGURE 6.27 Simplified three-vehicle ECP air brake model—implemented in Simulink.

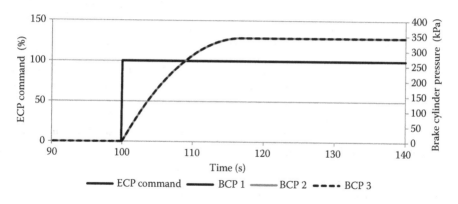

FIGURE 6.28   Simulation results—simplified train ECP brake model for 750 m brake pipe.

systems. It also presents perhaps the most challenges for modelling and simulation due to the non-linearities of the air gap (or coupler slack), draft gear spring characteristic (polymer or steel), and stick–slip friction provided by a wedge system. Due to these complexities, the common auto-coupler-friction-type draft gear wagon connection will be examined first. Other innovations such as slackless packages, drawbars and shared bogies are then more easily considered.

6.2.3.1   Conventional Auto-Couplers and Draft Gear Packages

A conventional auto-coupler and draft gear package is illustrated in the schematic in Figure 6.29. A schematic of the wedge arrangement of the draft gear unit is included in Figure 6.30. There are also several variations in wedge system designs, which are more complicated than Figure 6.30. Examples are shown in Figures 6.31 and 6.32.

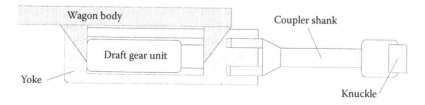

FIGURE 6.29   Conventional auto-coupler assembly.

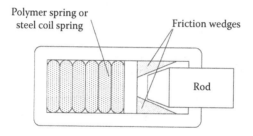

FIGURE 6.30   Friction-type draft gear unit.

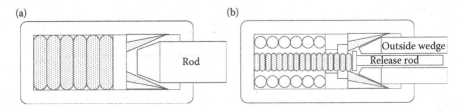

FIGURE 6.31 Variations on friction-type draft gear units. (a) Angled surfaces for increased wedge force, (b) release spring type.

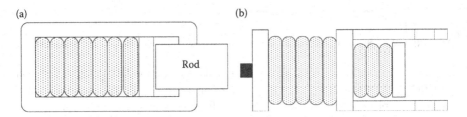

FIGURE 6.32 Polymer and elastomer draft gear units. (a) Polymer or elastomer only, (b) combined draft gear and yoke for differing buff and draft stiffness—polymer or elastomer springs.

When considering a wagon connection, two auto-coupler assemblies must be considered along with gap elements and also stiffness elements describing flexure in the wagon body. A wagon connection model will therefore appear as something similar to the schematic in Figure 6.33. Modelling the coupler slack is straightforward, using a simple dead zone. Modelling of the steel components including wagon body stiffness can be provided by a single linear stiffness. Work by Duncan and Webb [2] from test data measured on long unit trains identified cases where the draft gear wedges locked and slow sinusoidal vibration was observed. The behaviour was observed in distributed power trains when the train was in a single stress state. The train could be either in a tensile or compressed condition. The stiffness corresponding to the fundamental vibration mode observed was defined as the locked stiffness of the wagon connections. The locked stiffness for the freight trains tested was nominally a value

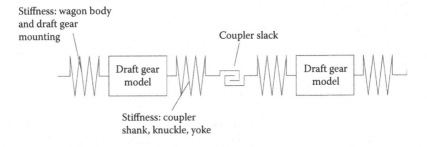

FIGURE 6.33 Components in a wagon connection model.

in the order of 80 MN/m [2]. As the locked stiffness is also the limiting stiffness of the system, it must be incorporated into the wagon connection model. The limiting stiffness is the sum of all the stiffnesses of the structural components and connections added in series, which includes components such as the coupler shank, knuckle, yoke, draft gear structure and the wagon body. It also includes any pseudo-linear stiffness due to gravity and bogie steering force components, whereby a longitudinal force is resisted by gravity as a wagon is lifted or forced higher on a curve. The limiting stiffness of a long train may therefore vary for different wagon loadings and on-track placement.

Wagon connection modelling can be simplified to a combined draft gear package model equivalent to two draft gear units, and including one spring element representing locked or limiting stiffness (see Figure 6.34).

Determination of the mathematical model for the draft gear model has received considerable attention in technical papers. For the purposes of providing a model for train simulation, a piecewise linear model representing the hysteresis in the draft gear friction wedge (or clutch) mechanism is usually used [2,13]. The problem of modelling the draft gear package has been approached in several ways. In early driver training simulators, when computing power was limited, it was common practice to further reduce the complexity of the dynamic system by lumping vehicle masses together and deriving equivalent connection models. As adequate computational capacities are now available, it is normal practice to model each wagon in detail [13,14]. It would seem reasonable in the first instance to base models on the hysteresis data published for the drop hammer tests of draft gear units. Typical draft gear response curves are shown in Figure 6.35.

The first thing to remember is that the published data as shown in Figure 6.35 represents the operating extreme simulated by a drop hammer test. The drop hammer of 12.27 tonne (27,000 lbs) impacts the draft gear at a velocity of 3.3 m/s, thus simulating an inter-wagon impact with a relative velocity between wagons of 6.6 m/s, (23.8 kph). In normal train operation, it would be hoped that such conditions are quite rare. Data recording of in-train forces of unit trains in both iron ore and coal haulage systems in Australia revealed that draft gear stiffness in normal operation could be very different from that predicted by drop hammer test data [2,13]. The approach taken by Duncan and Webb [2] was to fit a model to the experimental data as indicatively shown in Figure 6.36 using the piecewise linear functions.

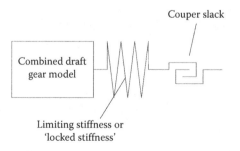

FIGURE 6.34   Simplified wagon connection model.

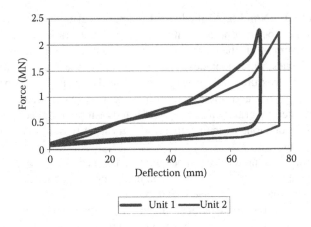

FIGURE 6.35 Typical published draft gear response data—drop hammer tests. (From C. Cole, in: *Handbook of Railway Vehicle Dynamics*, Chapter 9, Taylor & Francis, Boca Raton, FL, pp. 239–278, 2006.)

It will be noted that the model proposed by Duncan and Webb includes the locked stiffness as discussed earlier. A significant outcome from the train test data reflected in the model in Figure 6.36 was that unloading and loading could occur along the locked curve whenever the draft gear unit was locked. This cyclic unloading and loading could occur at any extension. Data from this train test program [2] and later work by [13] confirmed that the draft gear unit would remain locked until the force level reduced to a point close to the relaxation or unloading line. Owing to individual friction characteristics, there is considerable uncertainty about where 'unlocking' occurs. In some cases, unlocking was observed below the unloading curve, even sometimes reaching zero force.

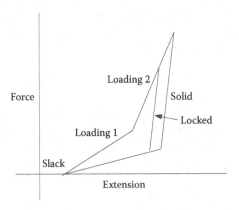

FIGURE 6.36 Typical piecewise linear wagon connection model [2].

Further refinement of wagon connection modelling was proposed by Cole in [13]. The difficulty presented by the work of Duncan and Webb [2] is that draft gear units, and so the mathematical models used to represent them, differ depending on the regime of train operation expected. Clearly, if extreme impacts were expected in simulation due to shunting or hump yard operations, a draft gear model representing drop hammer test data would be appropriate. Conversely, if normal train operations were expected, a wagon connection model of the form shown in Figure 6.36 would be appropriate. It was noted in [13] that the stiffness of the draft gear units for small deflections varied, typically 5–7 times the stiffness indicated by the drop hammer test data in Figure 6.35. It is therefore evident that, for mild inter-wagon dynamics (i.e. gradual loading of draft gear units), the static friction in the wedge assemblies can sometimes be large enough to keep draft gears locked. A model incorporating the wedge angles and static and dynamic friction was therefore proposed in [13] and published in detail in [1]. Results of this modelling approach are as shown in Figures 6.37 through 6.39.

Irrespective of the type of draft gear being used, the general principles for modelling are the same. They are generally characterised by a non-linear spring of some kind. Generally, a non-linear mathematical function or piecewise linear model will be required for the basic stiffness. The draft gear unit will have damping. As with many railway applications, damping provided by friction (Coulomb damping) is a popular choice as it can provide very high forces at slow velocities. Note that such properties are very hard to emulate in fluid- or polymer-based dampers. The damping in most freight draft gears is provided in both friction and polymer hysteresis. For passenger

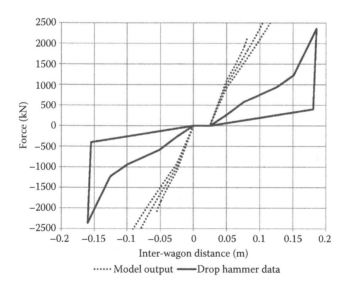

FIGURE 6.37 Wagon connection model response—slow loading (0.1 Hz). (From C. Cole, in: *Handbook of Railway Vehicle Dynamics*, Chapter 9, Taylor & Francis, Boca Raton, FL, pp. 239–278, 2006; C. Cole, *Proceedings of the Conference on Railway Engineering*, Rockhampton, Australia, 7–9 September 1998, pp. 187–194. With permission.)

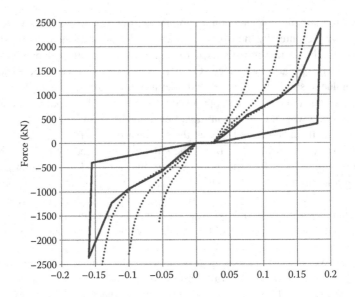

FIGURE 6.38 Wagon connection model response—mild impact loading (1 Hz). (From C. Cole, in: *Handbook of Railway Vehicle Dynamics*, Chapter 9, Taylor & Francis, Boca Raton, FL, pp. 239–278, 2006; C. Cole, *Proceedings of the Conference on Railway Engineering*, Rockhampton, Australia, 7–9 September 1998, pp. 187–194. With permission.)

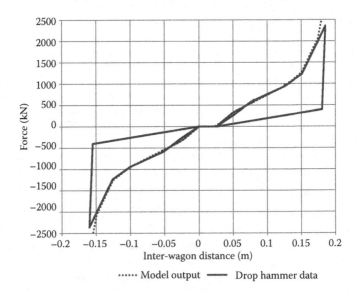

FIGURE 6.39 Wagon connection model response—shunt impact (10 Hz). (From C. Cole, in: *Handbook of Railway Vehicle Dynamics*, Chapter 9, Taylor & Francis, Boca Raton, FL, pp. 239–278, 2006; C. Cole, *Proceedings of the Conference on Railway Engineering*, Rockhampton, Australia, 7–9 September, 1998, pp. 187–194. With permission.)

vehicles, much softer draft gears are used and, in many cases, no friction elements are used. Examples of typical responses of these draft gears are shown in Figure 6.40. Note that the example of the passenger draft gear is simplified to a lookup table model with no velocity (load rate) dependence (Figure 6.40a). Polymer and elastomer elements can show both velocity and temperature dependence, but published work is still limited. Conversely, the studies in heavy haul railways have identified and allowed modelling of quite complex responses in friction draft gears (Figure 6.40b and [1]). Note that the stiffening of the freight-type draft gears in Figure 6.40b is due to the increase in friction to static levels as the draft gears slow and stop.

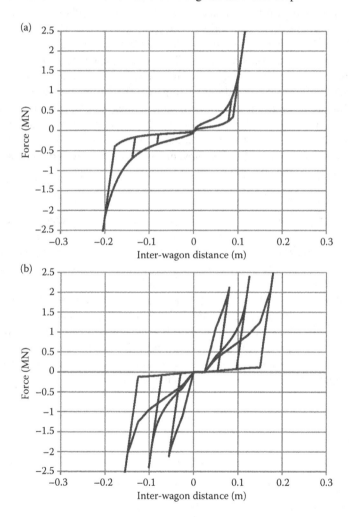

FIGURE 6.40 Typical responses—passenger and freight wagon draft gears (sinusoidal displacements are small, medium and large corresponding to the slow (0.1 Hz), medium (1.0 Hz) and fast (10.0 Hz) loading rates). (a) Passenger wagon-type draft gear (polymer or elastomer only), (b) freight wagon-type draft gear (friction wedge damping).

6.2.3.2 Slackless Packages

Slackless draft gear packages are sometimes used in bar-coupled wagons or integrated into shared bogie designs. The design of slackless packages is that the components are arranged to continually compensate for wear to ensure that small connection clearances do not get larger as the draft gear components wear. Slackless packages have been deployed in North American train configurations such as the trough train [15] and bulk product unit trains [16]. The advantage of slackless systems is found in reductions in longitudinal accelerations and impact forces of up to 96% and 86%, respectively, as reported in [15]. Disadvantages lie in the inflexibility of operating permanently coupled wagons and the reduced numbers of energy-absorbing draft gear units in the train. It is usual when using slackless coupled wagon sets that the auto-couplers at each end are equipped with heavier duty energy-absorbing draft gear units. The reduced capacity of these train configurations to absorb impacts can result in accelerated wagon body fatigue or even impact-related failures during shunting impacts. Modelling slackless couplings is simply a linear spring limited to a maximum stiffness appropriate to the coupling type, wagon body type and wagon loading. A linear or friction damper of very small value should be added to approximate small levels of damping available in the connection from friction in pins, movement in bolted or riveted plates and so on. See Figure 6.41 for the typical modelling set-up.

6.2.3.3 Drawbars

Drawbars refer to the use of a single link between draft gear packages in place of two auto-couplers. Drawbars can be used with either slackless or energy-absorbing draft gear packages. Early practice seems to favour retaining full capacity dry friction-type draft gear packages at the drawbar connections. The most recent practice in Australia is to utilise small short pack draft gear units at the drawbar connections. These short packs are quite stiff and provide only short compression displacements. They utilise only polymer or elastomer elements (no friction damping). A short pack would be a shorter version of the type of draft gear shown in Figure 6.32a. Modelling drawbars with energy-absorbing draft gear units are simply a matter of removing most of the coupler slack from the model as some slack will remain in pins and pocket components. A drawbar model schematic is shown in Figure 6.42. For cases

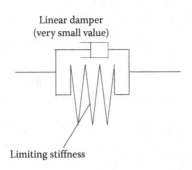

Linear damper
(very small value)

Limiting stiffness

FIGURE 6.41 Wagon connection model—slackless connection.

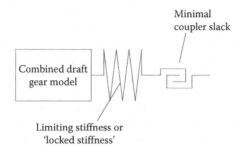

FIGURE 6.42   Wagon connection model—drawbar coupled wagon.

where short packs are used, then a short pack draft gear model replaces the traditional combined draft gear model.

6.2.4   TRAIN CONFIGURATIONS

Train configurations that are used in freight and heavy haul practice continue to evolve. There are three essential variables:

- The use of distributed power;
- The use of permanently coupled groups of wagons;
- The selection of brake control.

Several train marshalling and distributed power arrangements are given in Table 6.3. The type of configuration chosen will depend on many factors including the productive capacity required (route cycle times) and grades and curves. Faster haulage

TABLE 6.3
Some Examples of Train Configurations

Configuration Type	Diagram
Head end	
Head–tail	
Remote 2/3	
Remote mid-train	
Lead short rake	
Head mid-tail	
Remote 1/3, 2/3	

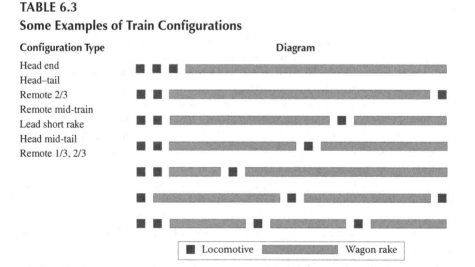

cycles will demand more locomotives be added as will steeper ruling grades. Sharper curves have the effect of requiring that in-train forces be limited favouring smaller locomotive groups and distributed power (see the six distributed power options in Table 6.3). Longer trains gain advantage from using permanently coupled wagon groups. There are smaller numbers of groups of eight and five used in freight and heavy haul trains. The advantage of grouped wagons is the improved longitudinal stability resulting from reduced coupling slack. The disadvantage is one for maintenance, as two or more wagons must be removed from the train to rectify a fault on one wagon. Tandem wagons are very common practice on Australian heavy haul trains, with some use of quad groups.

There are also trade-offs in the solution that can be chosen by considering the brake system. If distributed power is not required to limit in-train forces, particularly those at start up, then electronic braking could be a suitable solution that negates the need for distributed power. There are many cases where distributed power was originally adopted to ensure the reliable operation of the brake pipe and not because of traction forces.

6.2.5   Train Dynamics Model Development and Simulation

In previous sections, each part of the train dynamics system has been discussed and simple approaches to modelling introduced. These models can now be assembled to give an introduction to train dynamics issues with a simple three vehicle train. The train model is kept small to allow students and users of this text to easily observe the basic modelling and be able to repeat the modelling for themselves, giving a 'hands-on' experience of both modelling techniques and train simulation.

A model of a simple three vehicle train is shown in Figure 6.43, corresponding to Equations 6.1 through 6.3. Results are shown in Figure 6.44. The simulation has the same grades and curves and modelling as earlier examples. Note that the model is now so extensive that it is hard to read. A model showing just the linear connections in detail is given in Figure 6.45.

A significant issue in train dynamics is coupling free slack. While it adversely affects train dynamics and is the cause of impact forces, train-free slack has practical merits. The first and most obvious is that loose coupling systems will always have some slack. Even if slack is small, wear will always increase it. A second practical merit is that slack allows a staged application of force to the train. This has the advantage that relatively simple traction systems can be used. If a large train was tightly coupled, very precise traction control would be required as it would be necessary to move all the wagons at once. Conversely, for a train with slack, the locomotive firstly moves the first wagon or wagon group. Upon take up of the second connection gap, the locomotives and first wagon (or group) have momentum. The second wagon (or group) is therefore pulled by both adhesion and momentum. It still follows that better train dynamics can be achieved with less slack. The simulation given in Figure 6.46 has 20 mm slack in the connections, but in all other respects is identical to the train model used for the results in Figure 6.44. Note that impact force transients and impact accelerations exist at start up and at the change from tensile to compressive forces in Figure 6.46. As will be realised from the discussion

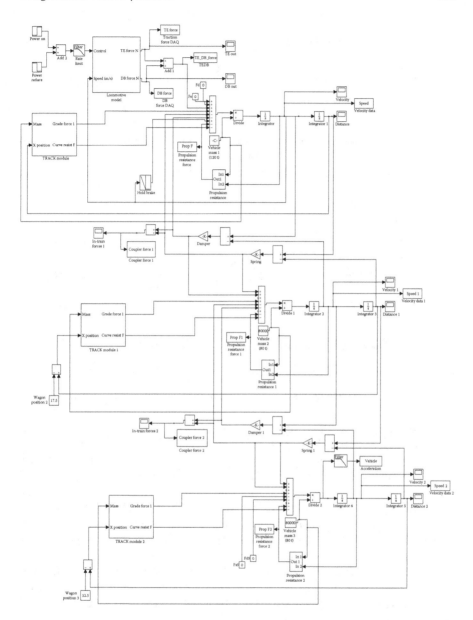

FIGURE 6.43 Three-vehicle train model with linear connections—implemented in Simulink.

on draft gear modelling, draft gears are usually not linear and provide considerable damping from friction on polymer properties. These factors together give the wagon connection a hysteresis characteristic with both stiffer and softer components. As discussed in [1], this complicated characteristic must also be combined with the stiffness of other train components. An example of a simulation with the same three

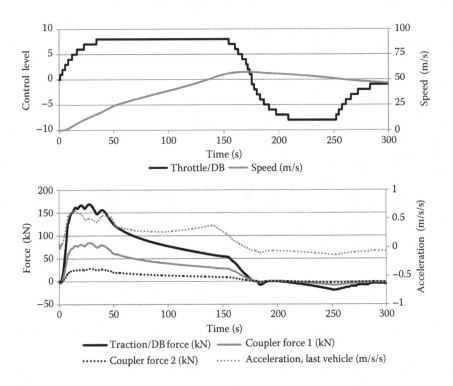

FIGURE 6.44 Simulation results—three-vehicle train model with linear connections.

mass model, but with simple upper and lower (loading and unloading) curves, is given in Figure 6.47, with cross plots of the draft gear responses in Figure 6.48.

It will be noticed that the effect of coupler slack in Figure 6.46 is to add impact forces and sudden changes in acceleration when compared to Figure 6.44. The train with linear connections in Figure 6.44 shows no impact where the coupler forces change from positive to negative. Note that the connection model used in Figure 6.45 is still a poor approximation of typical couplings as it has linear stiffness and viscous damping. To illustrate the modelling issues further, a very simple draft gear model is implemented in the train model to give the results shown in Figure 6.47; a cross plot of the model characteristic is shown in Figure 6.48. It should be noted that this model, although providing hysteresis indicative of Coulomb damping, is not giving representative results. In Figure 6.47, it will be noted that the plot of the second coupler force (the third force trace) is now noisy. This is caused by numerical instability in the solver, which is in-turn caused by the inadequacy of the model. Note that there is no solution to the question of the force state as the loading switches from 'loading' to 'unloading'. The consequence is that, between these curves, there is the impossibility of infinite stiffness. Some improvement can be achieved by adding viscous damping, but even if this is made large, there is still the case when a very, very slow velocity exists at the point where switching of direction occurs. The problem, of course, stems from the fact that the limiting stiffness (as discussed in Section 6.2.3) is ignored in the model. It is therefore reasonable to expect a much better result if

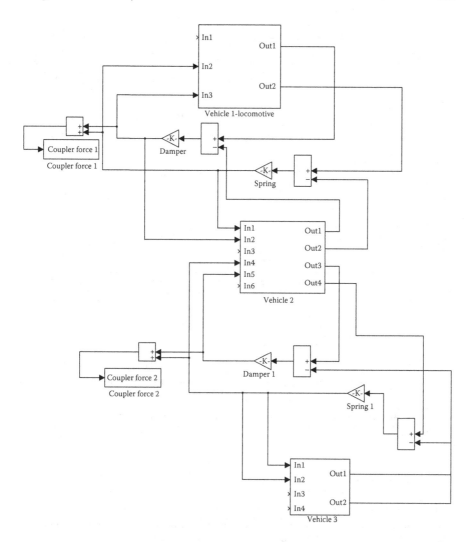

FIGURE 6.45   Three-vehicle train model—linear connections with vehicle details compressed to sub-systems—implemented in Simulink.

the model is more representative of reality; infinite stiffness does not exist in reality. In addition, infinite stiffness is problematic to step-wise numerical solvers, and it would be expected that a real model (with a finite maximum stiffness) will be much more easily solved. This in fact is what happens. More detailed models, accommodating limit stiffness, are implemented as shown in Figures 6.49 and 6.50. As the effect of limit stiffness can only be easily seen as a fundamental low vibration mode in a very long train, results from a long train model are shown in Figures 6.51 and 6.52. The effects of coupler slack are shown by the force transient peaks reaching 3 MN in the first case and 2.4 MN in the second case. Note that both of these correspond to changes in power control. Low-frequency vibration in both examples is then

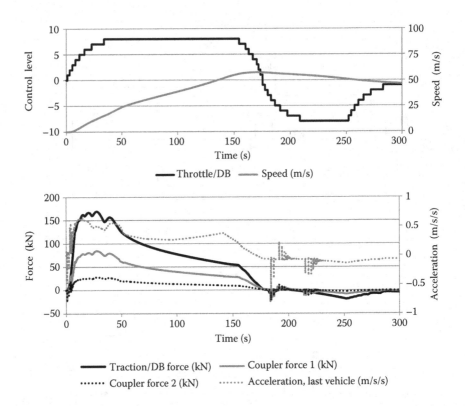

FIGURE 6.46 Simulation results—three-vehicle train model—linear connections with 20 mm coupling slack.

evidenced by the smooth sinusoids with very low damping, indicating the action of the wagon stiffness (limiting stiffness) rather than draft gear or damper movement.

The real freight train simulations in Figures 6.51 and 6.52 also illustrate the significant effect of train control inputs. As the dynamic system is large, changes to control can have significant effects. Traction control techniques, as shown by the examples in Figures 6.51 and 6.52, can significantly change the force results, hence the emphasis on specific driver training for various train types. In the case of heavy haul train systems, control strategies are focused on limiting forces to prevent component failure and ensure wagon stability. In passenger trains, different strategies focus on on-time running and passenger comfort. Note that the example given is for flat track and for a head end train. There will also be different practices for different situations of track topography and different train types. While power application is one issue, a more complicated issue is braking. There are two types of brakes. The first, known as dynamic brake, is a reverse traction force applied by the locomotive. Dynamic braking examples are shown in Figures 6.44, 6.46 and 6.47; the data after $t \cong 180$ s shows negative control levels for dynamic brake, negative acceleration and compressive coupler forces. As will be noted from the examples

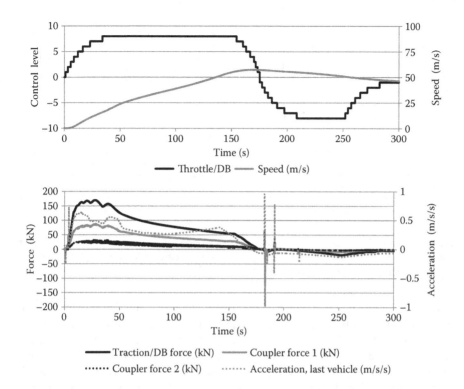

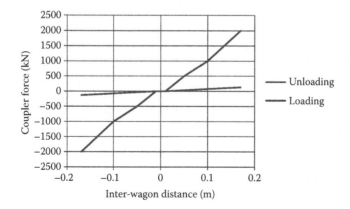

FIGURE 6.47 Simulation results—three-vehicle train model—non-linear connections with friction damping with 20 mm coupling slack.

FIGURE 6.48 Wagon connection model response—non-linear connection with estimated non-linear damping and 20 mm coupling slack.

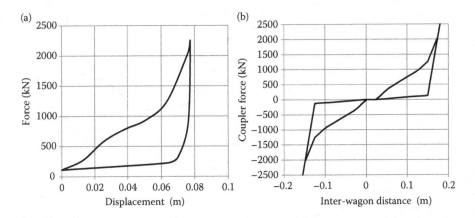

FIGURE 6.49 Wagon connection model response—non-linear connection with non-linear damping based on the drop hammer data, limiting stiffness and 20 mm coupling slack. (a) Drop hammer test data, (b) wagon connection model.

in Figures 6.7 and 6.8, dynamic braking usually delivers a smaller maximum force than the maximum tractive effort. An important difference is that locomotive systems are generally able to apply this force more quickly than traction can be applied. Another difference is that maximum traction is associated with very slow speeds, while maximum dynamic brake is available over a range of speeds, including much higher speeds. The second type of braking is the pneumatic train brake system, introduced earlier in Section 6.2.2.5 with modelling details and simulation results in Figures 6.23 through 6.28. It is important to realise that adding pneumatic braking to a train simulation is a co-simulation problem. The brake pipe model is a dynamic system in itself and will often require a different integrator step size. Computational

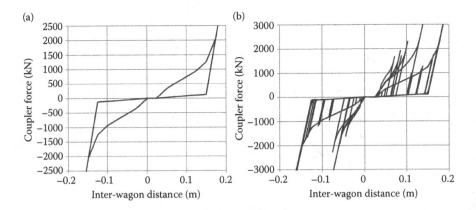

FIGURE 6.50 Wagon connection model response—non-linear connection with non-linear damping based on the drop hammer data, limiting stiffness, loading rate dependence and 20 mm coupling slack. (a) Wagon connection model response to different loading rates (single sine input), (b) wagon connection model response to different loading rates (single sine input).

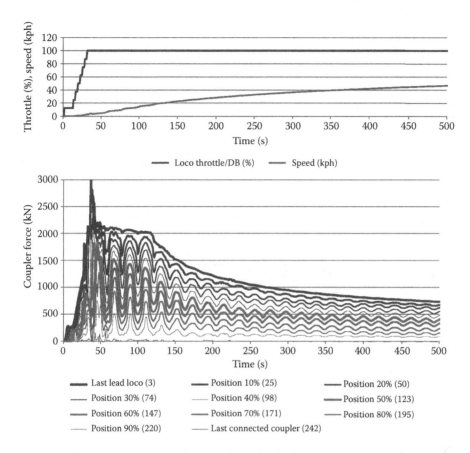

FIGURE 6.51 Simulation results—train start-up—243 vehicle train model (3 head end locomotives, 240 wagons)—non-linear connection with non-linear damping based on the drop hammer data and field data, limiting stiffness, loading rate dependence with 25 mm coupling slack between permanently coupled wagon pairs. Example: Fast throttle application.

fluid dynamic models of a pneumatic brake system are generally unworkable with a train simulator because of the very high computational demands. Brake pipe models in practical train simulators are usually partly or totally based on empirical equations. The brake pipe model implemented in Simulink in Figure 6.53 is taken from Figure 6.24 and uses standard signal processing blocks to approximate pipe behaviour. Gas laws are then used to model the cylinder fill. This model is implemented with the linear three-vehicle model to give the results in Figure 6.54. As the train is very short, pipe delays are minimal and, given linear connections, the coupler force results are well behaved. A more realistic simulation of over 115 km of train operation is given in Figures 6.55 and 6.56. It will be noted that both dynamic brake and mainly minimum brake pipe applications (50 kPa drop) are used to control the train speed. A zoom in of a braking event is shown in Figure 6.56. Sharper compressive force transients will be noted at $t \sim 3980$ s corresponding to the air brake application, while the triangular shape of the compressive force profile is corresponding to the

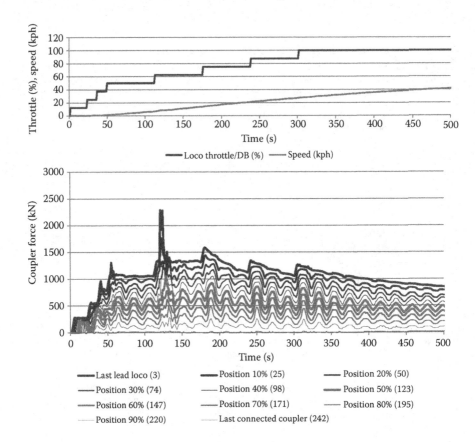

FIGURE 6.52 Simulation results—train start-up—243 vehicle train model (3 head end locomotives, 240 wagons)—non-linear connection with non-linear damping based on the drop hammer data and field data, limiting stiffness, loading rate dependence with 25 mm coupling slack between permanently coupled wagon pairs. Example: Train start up—throttle paused.

dynamic brake application. Note also that the limiting stiffness gives rise to small low-damped vibrations and that severe accelerations correspond to changes in control and switching from tensile to compressive coupler forces.

The simple non-linear model in Figure 6.48 does not include the locked or limiting stiffness as discussed earlier. A more complete modelling is shown in Figure 6.49.

6.3 INTERACTION OF LONGITUDINAL TRAIN AND LATERAL/VERTICAL WAGON DYNAMICS

The long tradition of analysing train dynamics and wagon dynamics separately is strongly entrenched in both software and standards. Train dynamics tends to be concerned only with longitudinal dynamics, while wagon dynamics tends to focus on just one vehicle (or a small number of vehicles) and on vertical and lateral dynamics. The assumption that coupler angles are so small that the consequential vertical and

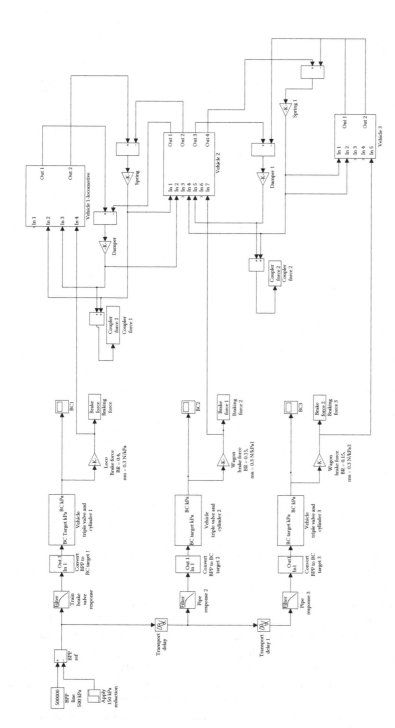

FIGURE 6.53 Three-vehicle train model—linear connections and brake system added (vehicle details compressed to sub-systems from Figure 6.45)—implemented in Simulink.

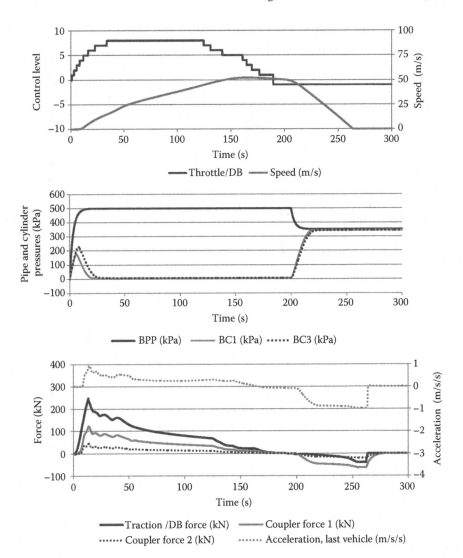

FIGURE 6.54 Simulation results—three-vehicle train model—linear connections with brake application.

lateral force components can be ignored does not necessarily hold as trains become heavier and longer and coupler forces become larger. Some possibilities for wagon instabilities were examined in [17], namely wheel unloading due to the lateral components of coupler forces and wagon lift due to mismatches in coupling height. In both these cases, the most severe events occur when an empty wagon is placed in a loaded train. It is evident that mechanisms of wagon instability can be more complex than just these clearly extreme cases. Further, wheel unloading can be added to by wagon body and bogie pitch induced by both track irregularities and train dynamics. An early paper is [6]. Just as there has been a questionable tradition of separating

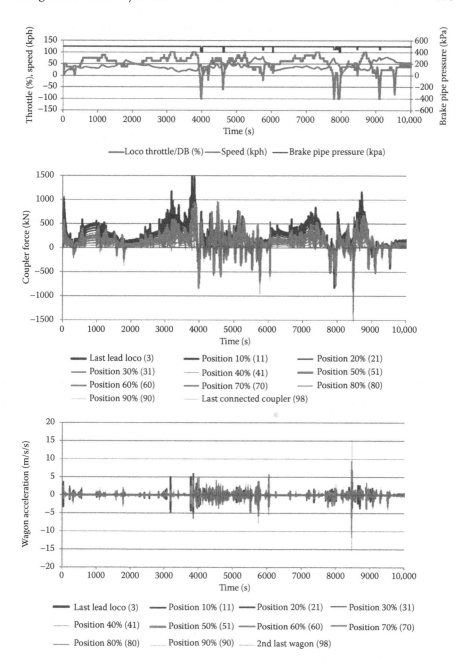

FIGURE 6.55 Overall simulation results—99 vehicle train model (3 head end locomotives, 96 wagons)—non-linear connection with non-linear damping based on the drop hammer data and field data, limiting stiffness, loading rate dependence with 25 mm coupling slack between permanently coupled wagon pairs. Example: 115.3 km train route simulation.

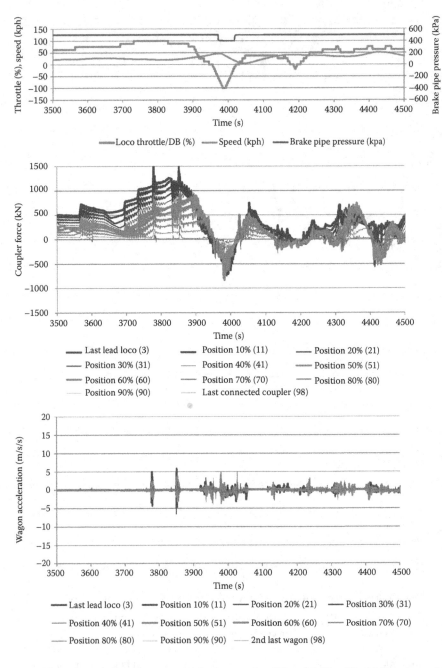

FIGURE 6.56 Overall simulation results—zoom-in of Figure 6.55—99 vehicle train model (3 head end locomotives, 96 wagons)—non-linear connection with non-linear damping based on the drop hammer data and field data, limiting stiffness, loading rate dependence with 25 mm coupling slack between permanently coupled wagon pairs. Example: 115.3 km train route simulation.

train and wagon dynamics, it would also be incorrect to discount the possibility of two or more mechanisms relating to longitudinal dynamics combining unfavourably.

6.3.1 WHEEL UNLOADING, WHEEL CLIMB AND ROLLOVER ON CURVES DUE TO LATERAL COMPONENTS OF COUPLER FORCES

For operating stability and true calculation of L/V ratios, it is important to combine the lateral coupler force components with individual wagon dynamics; but before this can be done, the lateral force components for the whole train simulation trip need to be calculated.

Coupler angles can be calculated by the equations provided in the AAR manual [18] or the easier technique developed by Simson [19] and utilised in many locomotive traction-steering studies [20,21]. This technique is easier to apply than the AAR method [18] and has no significant error penalty unless used for very sharp curves (error <0.1% at $R = 100$ m). The method also allows different curvatures to be applied for movement through transitions and is easier to implement in a train simulation context. The method uses the same assumptions as the AAR calculation, and makes the assumption that the two railway vehicles are coupled and are curving normally together ignoring any offset tracking and/or suspension misalignment at each bogie (the bogie pivot centre is assumed to be located centrally between the rails). The coupler pins are located at some distance overhanging the bogie centre distance. Figure 6.57a shows the configuration of the two vehicles with the angle between the two vehicles being θ and the angle of the coupler on vehicle 1 being ϕ. The angle ϕ of the coupler can be determined from the radius of curvature, the lengths between the adjacent wagon bogie centres and the overhang distance to the coupler pin and the coupler length. These define the angles α, β, γ, which are the cord angles to the arc for two vehicles and between the adjacent wagon bogie centres as shown in Figure 6.57a.

The relationship of θ, the angle between vehicle 1 and vehicle 2, and the cord arc angles is given and depicted in Figure 6.57a. The cord angle definitions are given as

$$\theta = \alpha + \beta + (2 * \gamma) \tag{6.18}$$

where

$$\alpha = \arcsin(BC_1/R_0); \quad \beta = \arcsin(BC_2/R_0); \quad \gamma = \arcsin(L/2/R_0)$$

where BC_i equals the half-length between bogie centres of vehicle i, L equals the cord length between the adjacent wagon bogie centres and R_0 is the radius of the track curve. By taking small-angle approximations, the above equations can be simplified to

$$\alpha = BC_1/R_0; \quad \beta = BC_2/R_0; \quad \gamma = L/2/R_0 \tag{6.19}$$

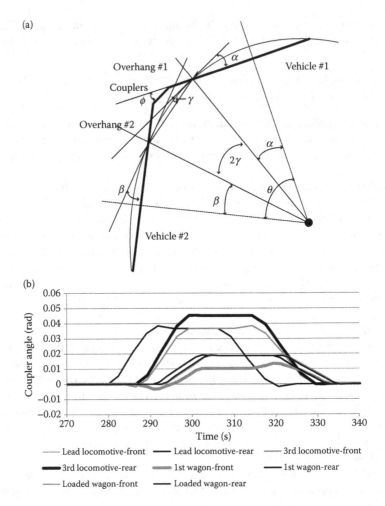

FIGURE 6.57 (a) Vehicle configuration during curving, (b) coupler angles for various vehicle combinations/locations.

As the arc of the curve must be common, it is not necessary to restrict this calculation to a single radius R_0, so the approach can be used to evaluate the coupler angles in transitions as follows:

$$\alpha = BC_1/R_{veh1}; \quad \beta = BC_2/R_{veh2}; \quad \gamma = L/2/R_L \tag{6.20}$$

Similarly, L can be approximated by using a small-angle assumption as

$$L = Ov_1 + Ov_2 + Cpl_1 + Cpl_2 \tag{6.21}$$

where Ov_i equals the overhang length of vehicle i (Ov_i = half the coupler pin centre-to-centre distance less half the bogie centre-to-centre distance) and Cpl_i equals the coupler length of vehicle i.

The coupler angle ϕ can be approximated by the equation:

$$\phi = (L * (\alpha + \gamma) - Ov_2 * \theta)/D \qquad (6.22)$$

where D is the combined length of the two couplers: $D = Cpl_1 + Cpl_2$

Coupler angles will differ for variations in vehicle length, overhang length and coupling length in the train. In heavy haul trains, the dimensions of wagons are more uniform and the dimensions can be standardised to just a few cases. In most heavy haul trains, just two vehicle lengths need to be analysed for locomotives and wagons as shown in Table 6.4. A few interesting observations can be made from Table 6.4:

- Increasing wagon length increases coupler angles;
- Increasing coupler length increases coupler angles;
- Unequal coupler pin distances from the bogie give large variations in coupler angles in long/short connections such as locomotive to wagon connections.

It is also interesting to note where these connection combinations might occur in a train.

Typical cases on curves (excluding transitions) for a head end train are:

- The lead locomotive—no coupling at the front; no lateral force at the front; equal coupling angle coupling at the rear if there are multiple locomotives; unequal coupling angles at the rear between the locomotive and the first wagon if there is only one locomotive;

TABLE 6.4
Coupler Angles for Various Vehicle Combinations on a 300 m Radius Curve

Dimensions[a]	Lead Vehicle			Trailing Vehicle			Angle on Lead Vehicle		
	B_1	C_1	Cpl_1	B_2	C_2	Cpl_2	Angle on	Rads	Deg
Datum									
Short wagon-short wagon	8.5	9.6	0.8	8.5	9.6	0.8	Wagon	0.0186	1.06
Locomotive-short wagon	16	21	0.8	8.5	9.6	0.8	Locomotive	0.0809	4.64
	8.5	9.6	0.8	16	21	0.8	Wagon	−0.0247	−1.41
Longer Wagons									
Long wagon-long wagon	10	13.4	0.8	10	13.4	0.8	Wagon	0.0250	1.43
Locomotive-long wagon	16	21	0.8	10	13.4	0.8	Locomotive	0.0651	3.73
	10	13.4	0.8	16	21	0.8	Wagon	−0.0025	−0.14
Longer Couplers									
Short wagon-short wagon	8.5	9.6	1.2	8.5	9.6	1.2	Wagon	0.0199	1.14
Locomotive-short wagon	16	21	1.2	8.5	9.6	1.2	Locomotive	0.0678	3.89
	8.5	9.6	1.2	16	21	0.8	Wagon	−0.0153	−0.88

[a] All dimensions in m. B is the bogie centre-to-centre distance, C is the coupler pin to pin distance. Cpl is the coupler length.

- The second and further locomotive—equal coupling angles at the front, between two locomotives; unequal coupling angles at the rear, between the locomotive and the first wagon;
- The first wagon—unequal coupling angles at the front, between locomotive and wagon; equal coupling angles at the rear, between identical wagons;
- In-train wagons—equal coupling angles front and rear, between identical wagons.

For a train with remote-controlled locomotives, the following cases are added:

- A single remote locomotive—unequal coupling angles at both the front and the rear, between the locomotive and the two connecting wagons;
- The lead locomotive in a remote group—unequal coupling angles between the locomotive and the wagon; equal coupling angles at the rear, between two locomotives;
- The single pusher locomotive in a remote group—unequal coupling angles at the front, between the locomotive and the wagon; no coupling at the rear; no lateral force at the rear.

Examples of angles from the various locomotive/wagon combinations are shown in Figure 6.57b. Where couplings are of 'like' vehicles, the angles are equal as expected. As angles are calculated as the vehicles move through the curve, a small over-throw 'kick' can be seen in all curves. When one vehicle has a longer bogie over-throw than another, for example, the last locomotive and the first wagon, larger and smaller angles than those on matching wagons can occur. In many configurations, the largest angle in the train occurs on the locomotive at the connection between the locomotive (or locomotive group) and the wagons. In such cases, the smallest angle occurs on the connecting wagon. If the mismatch is large enough, the wagon coupling can even be straight or opposite to the direction of the curve. As locomotives almost always have longer over-throw than wagons, the maximum coupler angle in the train is usually one of the locomotive to wagon connections. As the minimum wagon angle is also at this connection, the maximum lateral force components on wagons (which can be expected near locomotives) will actually occur at the connection between the first and second wagons in the rake, hence the second wagon usually has the greatest risk of overturning.

The methodology for calculation of coupler angles and associated forces is as follows:

- Calculate the front and rear coupler angles on all vehicles using the curvature data and wagon dimensions;
- This is completed using the equations above, but with the refinement of allowing changes in the overhang distances Ov_1 and Ov_2 in response to draft gear deflections as measured in the train simulations and allowing changes in the sum of $Cpl_1 + Cpl_2$ to incorporate the effect of coupling slack;
- Combine these angles with coupler forces to get lateral force components at the coupler pins.

This is done simply as

$$F_{lateral} = F_{coupler} * \phi \qquad (6.23)$$

- Use the moments to translate these forces to the bogies, noting that the forces are not equal during transitions. This parameter is designated as lateral forces from couplers.

$$F_{lfb} = [F_1 * (C + B) - F_2 * (C - B)]/(2B) \qquad (6.24)$$

$$F_{lrb} = [F_2 * (C + B) - F_1 * (C - B)]/(2B) \qquad (6.25)$$

where F_{lfb} and F_{lrb} are the lateral forces at the front and rear bogies, F_1 is the front lateral coupler force component, F_2 is the rear lateral coupler force component, C is the coupler pin half-distance and B is the bogie centre half-distance;
- Match the sign convention of longitudinal forces, considering the lateral forces as:
  - Positive if associated with tensile forces—these forces pull the wagon towards the centre of the curve (stringlining effects);
  - Negative if associated with compressive coupler forces—these forces push the wagon away from the centre of the curve (buckling effects).
- Add wagon centripetal forces to the lateral forces, assuming equal distribution of mass between front and rear bogies and using bogie curvature and superelevation. This parameter is designated as total quasi-static bogie lateral force.

$$F_{lfb\_TL} = m_w/2 \ (g * \sin(\psi) - V_w^2/abs(R) * \cos(\psi)) + F_{lfb} \qquad (6.26)$$

$$F_{lrb\_TL} = m_w/2 \ (g * \sin(\psi) - V_w^2/abs(R) * \cos(\psi)) + F_{lrb} \qquad (6.27)$$

where F_{lfb_TL} and F_{lrb_TL} are the total quasi-static lateral force at the front and rear bogies, m_w is the wagon mass, V_w is the wagon velocity, R is the curve radius and ψ is the track cant angle;
- Taking moments about each rail, the total quasi-static bogie lateral force could be used to calculate the quasi-static vertical forces on each side of each bogie, again assuming equal distribution of mass between front and rear bogies, and using bogie curvature and super-elevation. This parameter is designated as quasi-static bogie vertical force.

$$F_{vfhr\_TV} = m_w/2 * (g * \cos(\psi)/2 - g * \sin(\psi) * H_{cog}/d_c) + m_w/2 * V_w^2/abs(R)$$

$$* (\sin(\psi)/2 + \cos(\psi) * H_{cog}/d_c) - F_{lfb} * h_c/d_c \qquad (6.28)$$

$$F_{vflr\_TV} = m_w/2 * (g * \cos(\psi)/2 + g * \sin(\psi) * H_{cog}/d_c) + m_w/2 * V_w^2/abs(R)$$

$$* (\sin(\psi)/2 - \cos(\psi) * H_{cog}/d_c) + F_{lfb} * h_c/d_c \qquad (6.29)$$

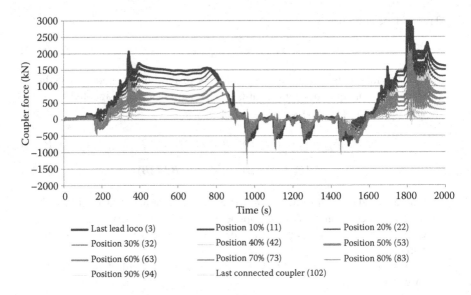

FIGURE 6.58 Simulation results—coupler forces–100 wagons, head end power.

$$F_{vrhr\_TV} = m_w/2 * (g * \cos(\psi)/2 - g * \sin(\psi) * H_{cog}/d_c) + m_w/2 * V_w^2/abs(R)$$

$$* (\sin(\psi)/2 + \cos(\psi) * H_{cog}/d_c) - F_{lrb} * h_c/d_c \tag{6.30}$$

$$F_{vrlr\_TV} = m_w/2 * (g * \cos(\psi)/2 + g * \sin(\psi) * H_{cog}/d_c) + m_w/2 * V_w^2/abs(R) *$$

$$* (\sin(\psi)/2 - \cos(\psi) * H_{cog}/d_c) + F_{lrb} * h_c/d_c \tag{6.31}$$

where F_{vfhr_TV} and F_{vflr_TV} are the total quasi-static vertical force at the front high and low rails, respectively, F_{vrhr_TV} and F_{vrlr_TV} are the total quasi-static vertical force at the rear high and low rails, respectively, H_{cog} is the height of the wagon centre of mass above the rail, d_c is the distance between wheel–rail contact points (~track gauge + 0.07 m) and h_c is the height of the wagon coupler above the rail.

Having derived total quasi-static bogie lateral force, it is also possible to calculate the quasi-static bogie L/V, but the vertical forces calculated cannot be used to give the bogie side L/V because the lateral force components cannot be separated into right and left rail components. To prevent confusion, it is not recommended that this parameter is used as it is very different from other definitions of L/V ratio.

To provide context for an example of the effects of coupler angles, a train simulation result is shown in Figure 6.58. Coupler angles, coupler lateral forces and lateral and vertical forces at the bogies are shown in Figures 6.59 through 6.62.

6.3.2 Wagon Body and Bogie Pitch Due to Coupler Impact Forces

Typically, for wagon dynamics studies, wagons are modelled as single vehicles with a longitudinal constraint. The models involve full modelling of the wheel–rail contact

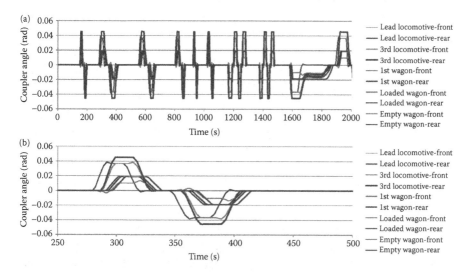

FIGURE 6.59   Simulation results. (a) Coupler angles—locomotive and wagon connections, (b) zoom in view.

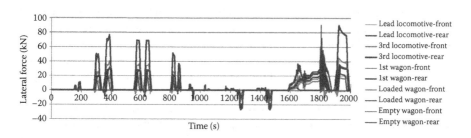

FIGURE 6.60   Simulation results—lateral forces due to couplers—locomotive and wagon connections (positive forces corresponding to tensile coupler stress).

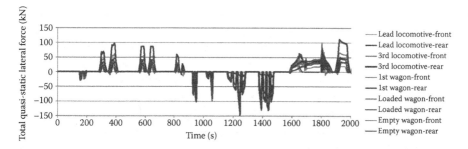

FIGURE 6.61   Simulation results—total quasi-static lateral forces—locomotives, loaded wagons and empty wagons (positive forces corresponding to force towards the low rail).

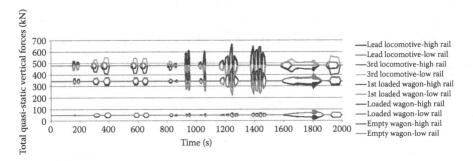

FIGURE 6.62 Simulation results—total quasi-static vertical forces—locomotives, loaded wagons and empty wagons.

patch and the 11 masses and up to 62 degrees of freedom (see Figure 6.63). Simulations can be obtained via available packages such as VAMPIRE, GENSYS and so on. For the consideration of wagon body and wagon bogie pitch, longitudinal forces and accelerations need to be known. A wagon model is also required that is computationally economical so that it can be completed for whole train trip simulations (e.g. >100 km). A simplified model is therefore desirable. The wagon pitch behaviour was modelled as in [17] with three pitch motions and three vertical motions (see Figure 6.64).

As only pitch and vertical motions are being modelled, the model can be further simplified by joining the bolster to the car body and modelling the bogie side frames and wheel sets as one mass. As some of the dynamic parameters are already calculated in the train simulation, the modelling of each wagon can be reduced to just six equations, three describing vertical motions of the wagon body and the two bogies and three describing the pitch rotations (see Figure 6.64a). It is also necessary to consider the effects of coupling heights and vertical force components between wagons. If wagons are the same type and load, these components will be small, but wagon bogie pitch motions will result in angles and vertical components. To ensure that correct interaction occurs at the couplings, three wagons are included (see Figure 6.64b), and only the results from the middle wagon are used. The other two wagons couple to points which are at a fixed height above the rail.

The modelling equations for the simplified model are reproduced from reference [17]:

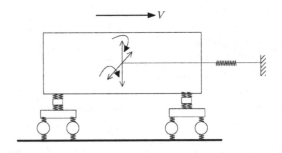

FIGURE 6.63 Schematic of a typical wagon dynamic model.

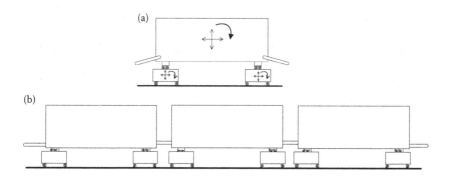

FIGURE 6.64 Simplified wagon pitch models implemented with longitudinal simulation. (a) Simplified single wagon pitch model, (b) simplified three wagon pitch model.

$$
\begin{aligned}
F_{zwb} = &-F_{c1} * (z_{wb} - z_n - C * \sigma_{wb})/(2C_{pl}) - F_{c2} * (z_{wb} - z_p + C * \sigma_{wb})/(2C_{pl}) \\
&+ F_{s1} + F_{s2} + F_{s3} + F_{s4} + F_{d1} + F_{d2} + F_{d3} + F_{d4} - abs(m_{wb} * g \cos(\lambda))
\end{aligned} \quad (6.32)
$$

$$
F_{zfb} = -F_{s1} - F_{s2} - F_{d1} - F_{d2} + F_{wrc1} + F_{wrc2} - m_{fb} * g \quad (6.33)
$$

$$
F_{zrb} = -F_{s3} - F_{s4} - F_{d3} - F_{d4} + F_{wrc3} + F_{wrc4} - m_{rb} * g \quad (6.34)
$$

$$
\begin{aligned}
M_{wb} = &-F_{s1} * (B + l_s) - F_{s2} * (B - l_s) + F_{s3} * (B - l_s) + F_{s4} * (B + l_s) \\
&- F_{d1} * (B + l_d) - F_{d2} * (B - l_d) + F_{d3} * (B - l_d) + F_{d4} * (B + l_d) \\
&+ F_{c1} * (z_{wb} - z_n - C * \sigma_{wb})/(2C_{pl}) * C - F_{c2} * (z_{wb} - z_p + C * \sigma_{wb})/(2C_{pl}) * C \\
&- F_{c1} * (h_{cg} + C * \sigma_{wb}) + F_{c2} * (h_{cg} - C * \sigma_{wb}) + m_{fb} * h_{cg} * a_w \\
&+ m_{rb} * h_{cg} * a_w + h_{cg} * (m_w + m_{fb} + m_{rb}) * g * \sin(\lambda)
\end{aligned} \quad (6.35)
$$

Note that the moment from the longitudinal reaction at the centre bowl connection is calculated from the bogie inertia term $m_{fb} * h_{cg} * a_w + m_{rb} * h_{cg} * a_w$. This is also done in Equations 6.36 and 6.37 resulting in

$$
\begin{aligned}
M_{fb} = &-F_{wrc1} * A + F_{wrc2} * A - F_{s1} * l_s + F_{s2} * l_s - F_{d1} * l_d + F_{d2} * l_d \\
&+ m_{fb} * h_b * a_w + F_{brake}/R_w
\end{aligned} \quad (6.36)
$$

$$
\begin{aligned}
M_{rb} = &-F_{wrc3} * A + F_{wrc4} * A - F_{s3} * l_s + F_{s4} * l_s - F_{d3} * l_d + F_{d4} * l_d \\
&+ m_{rb} * h_b * a_w + F_{brake}/R_w
\end{aligned} \quad (6.37)
$$

where:
A is the axle centre half-length;
B is the bogie centre half-length;
C is the coupler pin centre half-length;
Cpl is the coupler length;
F_{zwb}, F_{zfb} and F_{zrb} are the vertical force on the wagon body, front bogie and rear bogie, respectively;
F_{c1} and F_{c2} are the front and rear coupler force, respectively;
F_{s1} and F_{s2} are the spring force from the front and rear halves of the two spring nests in the front bogie, respectively;

F_{s3} and F_{s4} are the spring force from the front and rear halves of the two spring nests in the rear bogie, respectively;

F_{d1} and F_{d2} are the damper force from the front and rear wedges in the front bogie, respectively;

F_{d3} and F_{d4} are the damper force from the front and rear wedges in the rear bogie, respectively;

F_{wrc1}, F_{wrc2}, F_{wrc3} and F_{wrc4} are the total vertical wheel–rail contact force per axle on wheel sets 1, 2, 3 and 4, respectively;

F_{brake} is the bogie braking force;

H_b is the height of the coupling line above the bogie CoG;

M_{wb}, M_{fb} and M_{rb} are the moment about the pitch axis on the wagon body, front bogie and rear bogie, respectively;

R_w is the wheel radius;

a_w is the longitudinal acceleration of the wagon obtained from the train simulation;

h_{cg} is the height of wagon body CoG above coupling line;

l_s is the distance to force centroid of spring half-nest;

l_d is the distance to line of action of wedge dampers;

m_{wb} is the mass of the wagon body;

m_{fb} is the mass of the front bogie;

m_{rb} is the mass of the rear bogie;

z_{wb} is the height of the centre of mass of the wagon body;

z_n is the height of the centre of mass of the wagon body connecting to the front;

z_p is the height of the centre of mass of the wagon body connecting to the rear;

σ_{wb} is the pitch angle of the wagon body;

σ_{fb} is the pitch angle of the front bogie;

σ_{rb} is the pitch angle of the rear bogie;

λ is the track grade angle.

To provide context for an example of wagon body pitch, a train simulation result is shown in Figure 6.65. A hypothetical heavy haul train is simulated with all wagons loaded. To induce wagon body pitch, a minimum brake application is applied. Compressive coupler forces are induced as shown in Figure 6.66. Details of the coupler forces at the seventh wagon are shown in Figure 6.67, axle forces in Figure 6.68 and 'zoom-in' on axle forces showing body pitch in Figure 6.69.

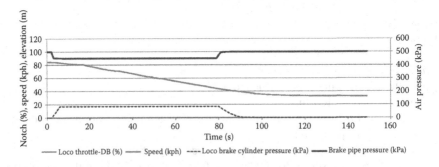

FIGURE 6.65 Simulation results—operational data—loaded train.

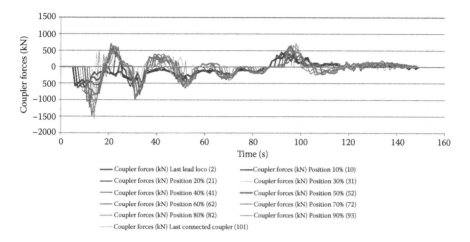

FIGURE 6.66   Simulation results—coupler forces—loaded train.

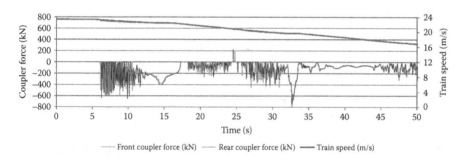

FIGURE 6.67   Simulation results—selected coupler force and operational data, wagon #7—loaded train.

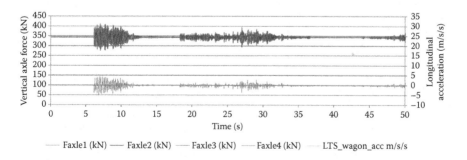

FIGURE 6.68   Simulation results—selected axle force data, wagon #7—loaded train.

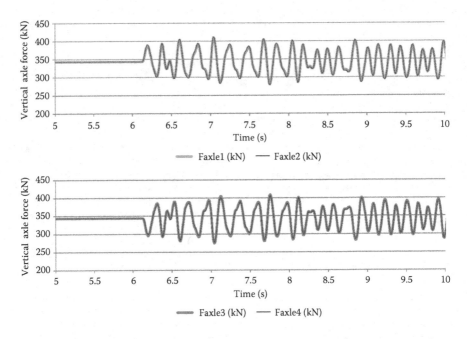

FIGURE 6.69   Simulation results—zoom-in on axle force data, wagon #7—loaded train.

Similarly, a hypothetical heavy haul train is simulated with all wagons empty. To induce wagon bogie pitch, a minimum brake application is applied. Details of the coupler forces at the seventh wagon are shown in Figure 6.70, axle forces in Figure 6.71 and 'zoom-in' on axle forces showing body pitch in Figure 6.72.

6.3.3 WAGON LIFTOFF DUE TO VERTICAL COMPONENTS OF COUPLER FORCES

Wagon lift can more easily occur if there is mismatch in coupling heights. The more severe case for vertical force components from coupling wagons with different coupling heights, either empty/loaded combinations or wagons of different type, can

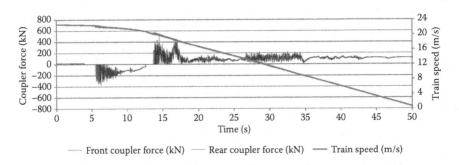

FIGURE 6.70   Simulation results—selected coupler force and operational data, wagon #7—empty train.

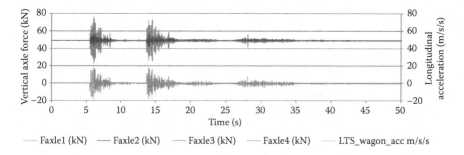

FIGURE 6.71   Simulation results—selected axle force data, wagon #7—empty train.

also be handled by the three-model approach. A schematic of this case is shown in Figure 6.73. It is assumed that the effect of a slight pitch angle on the adjacent wagons will have no significant effect on the wagon under study.

To provide context for examples of wagon lift off instability, train simulation results for coupler forces are shown in Figures 6.74 and 6.76. Figure 6.74 involves coupler tension, so the empty wagon is effectively pulled downward by the couplers, increasing wheel loads as shown in Figure 6.75. This situation increases wagon stability. The second example in Figures 6.76 and 6.77 is the opposite, involving coupler compression. In this case the wagon is lifted off the track by the couplers, and severe wheel unloading occurs as shown in Figure 6.77. If such events are severe enough, complete wheel lift off and consequent jack knifing can occur.

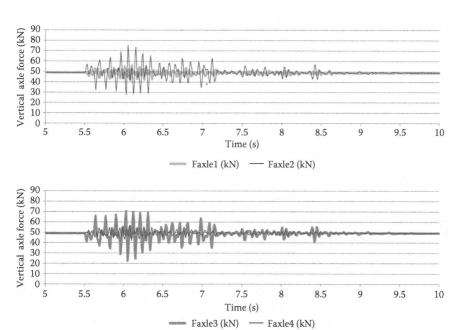

FIGURE 6.72   Simulation results—zoom-in on axle force data, wagon #7—empty train.

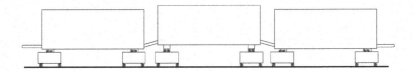

FIGURE 6.73   Simplified wagon pitch model implemented as a three-wagon model.

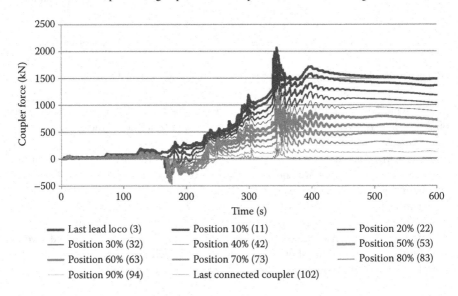

━━ Last lead loco (3)	━━ Position 10% (11)	━━ Position 20% (22)
── Position 30% (32)	┈┈ Position 40% (42)	▬▬ Position 50% (53)
━━ Position 60% (63)	── Position 70% (73)	── Position 80% (83)
┈ Position 90% (94)	── Last connected coupler (102)	

FIGURE 6.74   Simulation results—coupler forces—loaded train, traction case.

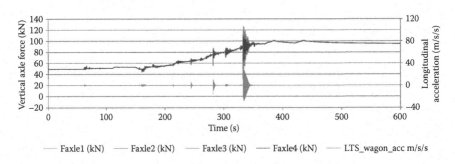

── Faxle1 (kN)	── Faxle2 (kN)	── Faxle3 (kN)	── Faxle4 (kN)	── LTS_wagon_acc m/s/s

FIGURE 6.75   Simulation results—selected axle force and acceleration data, empty wagon #7—loaded train, traction case.

6.4   LONGITUDINAL COMFORT

Ride comfort measurement and evaluation is often focussed on accelerations in the vertical and lateral directions. The nature of longitudinal dynamics is that trains are only capable of quite low, steady accelerations and decelerations due to the limits imposed by adhesion at the wheel–rail interface. Cleary, the highest steady acceleration achievable will be that of a single locomotive, giving a possible ~0.3 g assuming

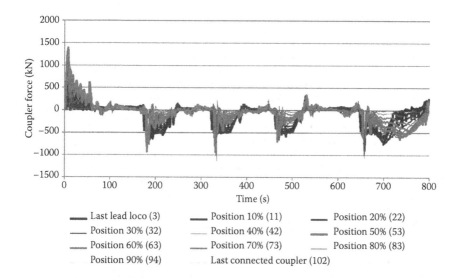

FIGURE 6.76 Simulation results—coupler forces—loaded train, braking case.

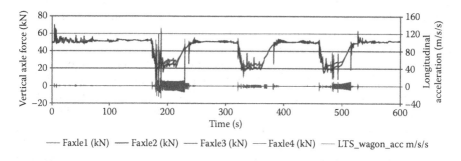

FIGURE 6.77 Simulation results—selected axle force and acceleration data, empty wagon #7—loaded train, braking case.

30% wheel–rail adhesion and driving all wheels. Typical train accelerations are of course much lower, of the order 0.1–1.0 m/s² [12]. Braking also has the same adhesion limit, but rates are limited to values much lower to prevent wheel locking and wheel flats. Typical train deceleration rates are of the order 0.1–0.6 m/s [12]. The higher values of acceleration and deceleration in the ranges quoted correspond to passenger and suburban trains. The only accelerations that contribute to passenger discomfort or freight damage arise from coupler impact transients. These events are of an irregular nature, so frequency spectral analysis and the development of ride indices are inappropriate in many instances. It is more appropriate to examine the maximum magnitudes of single impact events. A recent survey of literature and standards found no explicit standard for longitudinal ride comfort in relation to longitudinal impacts. The most recent standards [22,23] focus on an aggregate of comfort parameters. These methods typically use root mean square values or statistical methods such as described in ISO 10056, which takes the 95th percentile of

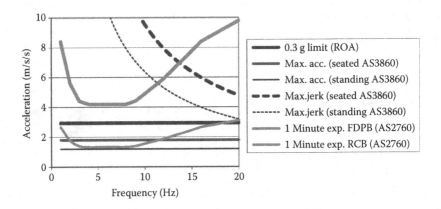

FIGURE 6.78   Passenger comfort acceleration limits.

weighted root mean square values over intervals of 5 s. It should also be noted that, with the move to multiple unit suburban trains (rather than mainline passenger operation), vehicles have distributed power, blended and/or electronic braking, advanced slip controls and permanently coupled cars. It follows that, for these applications, longitudinal dynamics may not be such an issue. The same could be stated for high speed trains. In such cases, if longitudinal comfort is considered at all, it will be considered in the aggregate of comfort parameters.

There is also a difference in comfort standards depending on the orientation of the human body in the train. Several standards, including the one adopted in the Sydney Suburban system in Australia, apply the lateral comfort limits to longitudinal accelerations if the passengers are orientated on longitudinal seating, rather than the traditional lateral seating, see RDS 7513.3, RISSB, Australia [24]. It therefore remains that standards for longitudinal comfort must still rely on more general and non-railway specific standards such as AS2670 [25] and AS3860 [26]. The maximum acceleration limits specified by these standards are

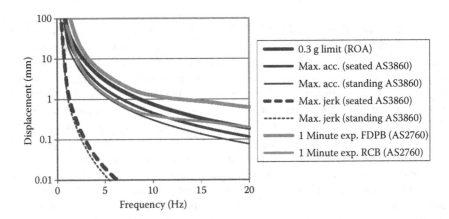

FIGURE 6.79   Passenger comfort displacement limits.

plotted in Figure 6.78, with associated displacements in Figure 6.79. A longer discussion can be found in [1].

6.5   ENERGY CONSIDERATIONS

Minimisation of energy usage is often a popular emphasis in train management. It is helpful to examine the way energy is utilised before innovations or changes to practice are adopted. Air resistance for example is often over-stated. A breakdown of the Davis equation [11] shows the significance of air resistance compared to curving resistance and rolling resistance factors and grades (see Figure 6.80). It will be noticed that a 1 in 400 grade, or 0.25%, is approximately equal to the propulsion resistance at 80 kph.

The minimum energy required for a trip can be estimated by assuming an average train speed and computing the sum of the resistances to motion, and not forgetting the potential energy effects of changes in altitude. The work done to get the train up to running speed once must also be added. As the train must stop at least once, this energy is lost at least once. Any further energy consumed will be due to signalling conditions, braking, stop-starts and the design of grades. Minimum trip energy can be estimated as

$$E_{min} = \frac{1}{2}m_t v^2 + m_t gh + \sum_{i-1}^{q}\left(m_i * \sum_{j=1}^{r}\left[\int_{0}^{x=lcj} F_{crj} \, . \, dx \right] \right) + \sum_{i=1}^{q}\left(m_i * \int_{0}^{x=L} F_{prj} \, . \, dx \right)$$

(6.38)

where E_{min} is the minimum energy consumed, J; g is the gravitational acceleration in m/s^2; h is the net altitude change, m; L is the track route length, m; m_i is the individual vehicle mass i, kg; m_t is the total train mass, kg; F_{crj} is the curving resistance

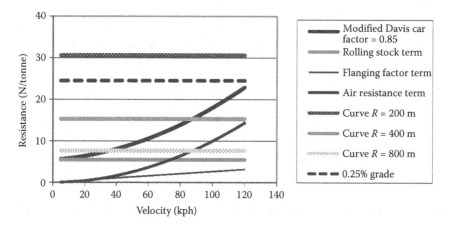

FIGURE 6.80   Comparative effects of resistances to motion. (From C. Cole, in: *Handbook of Railway Vehicle Dynamics*, Chapter 9, Taylor & Francis, Boca Raton, FL, pp. 239–278, 2006.)

TABLE 6.5

Energy Losses Equivalent to One Train Stop for a Train Running at 80 kph

Energy Parameter	Equivalent Loss	Units
Gravitational potential energy (2nd term, Equation 6.38)	~25	Metres of altitude
Curving resistance (3rd term, Equation 6.38)	~16	Kilometres of curving resistance on $R = 400$ m curve
Propulsion resistance (4th term, Equation 6.38)	~18	Kilometres of propulsion resistance
Air resistance (part of propulsion resistance)	~36	Kilometres of air resistance

for curve j in Newtons; F_{pri} is the propulsion resistance for vehicle i in Newtons; q is the number of vehicles and r is the number of curves.

Unless the track is extremely flat and signalling conditions particularly favourable, the energy used will be much larger than given by the above equation. It is however a useful equation in determining how much scope exists for improved system design and practice. It is illustrative to consider a simple example of a 2000 tonne freight train with a running speed of 80 kph. The work done to bring the train to speed, represented in Equation 6.38 by the kinetic energy term, is lost every time the train must be stopped and partly lost by any brake application. The energy loss per train stop in terms of other parameters in Equation 6.38 is given in Table 6.5.

What can be seen at a glance from Table 6.5 is the very high cost of stops and starts compared to other parameters. Air resistance becomes more significant for higher running speeds. High densities of tight curves can also add considerable costs. It should be noted that this analysis does not include the additional costs in rail wear or speed restriction also caused by curves.

In negotiating crests and dips, the driver has the objectives of minimising the power loss in braking and managing in-train forces. In approaching the top of a crest, at some point close to the top (depending on grades, train size, etc.), power should be reduced to allow the upgrade to reduce the train speed, the objective being that excess speed requiring severe braking will not occur as the train travels down the next grade. Similarly, when negotiating dips, braking should be reduced at some point while approaching the dip to allow the current falling grade to 'push' the train to reduce the power needed in climbing the next rising grade. It can be seen that there is considerable room for variations in judgement and hence variation in energy usage. Work published in [7] indicated variations in fuel usage of up to 42% due primarily to differences in the way drivers manage the momentum of trains.

6.6   TRAIN CONTROL, MANAGEMENT AND DRIVING PRACTICES

Train management and driving practices have received considerable attention in literature dating back several decades. Technology developments such as the transitions from steam to diesel and electric locomotives, improved locomotive traction control systems, remote control locomotives, operation of very long heavy haul unit trains and the operation of high-speed passenger services have ensured that this area continues to evolve. Train management and driving practices will differ for different rail operations, and optimum practice depends on the different targets. Train dynamics management has implications for the following parameters:

- Production:
  - On-time running;
  - Tonnes transported.
- Energy.
- Asset management:
  - Safety:
    - Speed compliance;
    - Failure prevention;
    - Wagon stability.
  - Equipment:
    - Failure prevention;
    - Fatigue life of components.

Suburban train drivers will be motivated primarily by the need to run to time. A secondary consideration may be energy consumption. Longitudinal dynamics will have minimal consideration as cars are connected with minimal slack and cars will usually have distributed traction and slip controls for both traction and braking. Passenger express services will be similarly motivated. Slow passenger services with locomotive hauled passenger cars will share the concerns of running to time with the next priority being the smoothness of passenger ride. Passenger train driver practices in times past have often included energy consumptive 'power braking' to minimise the slack action. The technique of 'power braking' is where locomotive power is left on and locomotive brakes not used while low levels of pneumatic brake are used. The idea is that wagon impacts due to brake pipe delays will be less severe. Power braking is also used to reduce the impacts in normal running over undulations.

Mixed freight train practice, while not motivated by passenger comfort, will share some similar driving techniques to ensure the train stability. This is particularly the case when trains are operated with mixes of empty and loaded wagons. Running to time will be an emphasis on some systems depending on the type of freight. Differing from passenger systems, energy consumption is a significant freight cost factor and is emphasised in freight operations. The operation of bulk product/uniform module-type freight trains (unit trains or block trains, e.g. carrying minerals, grain or containers) can be optimised to the specific source/destination requirements. In some cases, timeliness is a secondary concern while tonnage per week targets must be achieved.

Heavy haul trains are characterised by longer and heavier trains of a single wagon and payload type. Again, on time running is usually a secondary consideration to tonnage per week targets. As the trains can be large (up to 50,000 tonnes), operations place a high emphasis on train dynamics as these systems seek high productivity and safety targets. Note that a derailment of a large heavy haul train can easily cost in excess of $20 million (USD) at the time of writing (2014). This threat must be managed by clearly understanding the risks associated with:

- Coupler and draft gear assembly failure (Section 6.2.3) by reaching yield and ultimate stress;
- Wagon stability (Section 6.3);
- Failure of components due to fatigue.

As in-train forces in current heavy haul trains in Australian service can easily exceed the tensile yield of couplers as designated in the AAR Standard M-211 [27] of 1.8 MN, it is normal practice to use train simulation at both design and implementation stages of heavy haul train projects. Control of in-train forces is then the responsibility of train driver practice unless a system of automated train control (driverless) is adopted. Safe operation of heavy haul trains is therefore highly dependent on adequate driver training programs and compliance with recommended practices. Wagon stability is a further complication and arises in different ways. When new heavy haul routes are designed and built, wagon stability issues tend to be designed out by keeping the radius of curves adequately large and grades adequately small for the trains concerned. When this occurs, quite extraordinary designs can be achieved; for example, heavy haul train systems have been built with wagon to locomotive mass ratios as high as 90. Wagon stability becomes an issue where trains are made larger but are operating on older infrastructure. The situation arises that larger in-train forces are applied on sharper curves. In addition, grade forces are larger. Risks for wagon stability that need to be understood and managed include:

- In-train forces at start up and braking in empty trains on curves (wagon overturning);
- Longitudinal wagon accelerations at start up and braking in empty trains (wagon pitch);
- The above two scenarios combined;
- Operation of empty wagons in loaded trains:
  - Lateral instability on curves—overturning and wheel climb;
  - Wagon lift off and jack knifing.

Despite the differences in operation, a common threat to train dynamics management is the issue of speed control and hence management of train momentum. For suburban passenger trains, speed must be managed to ensure timeliness and adequate stopping distances for signals and for positioning at platforms. For longer locomotive hauled passenger, freight and heavy haul unit trains, the problem of momentum control becomes even more significant due to the larger masses involved. In general, it is desirable to apply power gradually until in-train slack is taken up. During running, it is desirable to minimise the braking and energy wastage, utilising coasting where possible. Route running times will limit the amount of time that the train can coast. Longer trains can coast over undulating track more easily than shorter trains due to the grade forces being partially balanced within the train length. Stopping is achieved at several different rates. Speed can be reduced by removing power and utilising rolling resistance (slowest), application of dynamic braking, application of minimum pneumatic braking, service application of pneumatic braking and emergency application of pneumatic braking (fastest). The listed braking methods are also in order of increasing energy wastage and increased maintenance costs. The selection of and blending of freight train braking is quite complicated and will often be governed by practice rules. Note that the recent adoption of electronically controlled brakes is expected to simplify and revolutionise braking practice, but traditional train braking, controlled by the brake pipe, will also exist for many years to come. Braking scenarios are listed below:

- Locomotive dynamic braking only;
- Locomotive air braking only (usually forbidden);
- Minimum braking with locomotive brakes off (applying a 30% brake application to wagons only);
- Minimum braking with locomotive brakes on (applying a 30% brake application to all vehicles);
- Service braking with locomotive brakes off (applying >30% up to 100% brake application to wagons only);
- Service braking with locomotive brakes on (applying >30% up to 100% brake application to all vehicles);
- Emergency braking (100% brake application to all vehicles, braking at a higher pressure on AAR (USA) systems);
- Penalty braking (100% brake application to all vehicles, braking at a higher pressure on AAR (USA) systems).

Policies differ as to whether dynamic braking is allowed during an air braking application or not. Much of braking practice is dependent on train type. Heavy haul trains with very long wagon rakes will tend to favour only very mild applications of dynamic brakes and greater use of minimum applications. The use of distributed power (remote locomotives) is often driven by the need to improve brake control, which is achieved by controlling the brake pipe at multiple points and so reducing application delays and inter-wagon impacts. The use of electronically controlled pneumatic brakes (ECP) is now an option for greatly improving braking, as all wagons can be braked simultaneously. If locomotive brakes are adjusted appropriately, braking can be tuned so that there are no in-train forces or inter-wagon impacts at all. In other differences, the ECP systems provide greater selectivity allowing application in the range 10–100% and permitting graduated adjustment during both the application (increasing the brake forces) and release (reducing the braking forces). Note that traditional braking systems do not all have 'graduated release'. Graduated release exists in Europe, but not in Australian and North American systems.

6.7   DESIGN CONSIDERATIONS

As shown in this chapter, the study of longitudinal dynamics has significant implications for design strength of coupling and draft gear components. Heavy haul trains continue to push the limits of train design in both yield strength and fatigue life.

In addition, the study of the interaction of longitudinal train dynamics and wagon dynamics reveals a much more complex area that needs design consideration. The tendency towards shorter couplings and shorter over-throw lengths give favourable reductions to lateral forces on curves. Conversely, shorter couplers increase the risk of wagon lift and jack knifing of empty wagons in otherwise loaded trains. Large longitudinal wagon accelerations increase the risk of wagon body pitch and bogie pitch. The use of permanently coupled wagon pairs, quads or eights with reduced inter-wagon slack is a suitable method of reducing slack action and wagon body/ bogie instability. The study of longitudinal train dynamics is therefore a valuable design tool both for train and track corridor design in addition to its traditional driver training applications.

REFERENCES

1. C. Cole, Longitudinal train dynamics, in *Handbook of Railway Vehicle Dynamics*, S. Iwnicki (Ed.), Chapter 9, Taylor & Francis, Boca Raton, FL, pp. 239–278, 2006.
2. I.B. Duncan, P.A. Webb, The longitudinal behaviour of heavy haul trains using remote locomotives, *Proceedings of the 4th International Heavy Haul Conference*, Brisbane, Australia, 11–15 September 1989, pp. 587–590.
3. B.J. Jolly, B.G. Sismey, Doubling the length of coals trains in the Hunter Valley, *Proceedings of the 4th International Heavy Haul Conference*, Brisbane, Australia, 11–15 September 1989, pp. 579–583.
4. R.D. Van Der Meulen, Development of train handling techniques for 200 car trains on the Ermelo–Richards Bay line, *Proceedings of the 4th International Heavy Haul Conference*, Brisbane, Australia, 11–15 September 1989, pp. 574–578.
5. M. El-Sibaie, Recent advancements in buff and draft testing techniques, *Proceedings of the 5th International Heavy Haul Conference*, Beijing, China, 6–11 June 1993, pp. 146–150.
6. M. McClanachan, C. Cole, D. Roach, B. Scown, An investigation of the effect of bogie and wagon pitch associated with longitudinal train dynamics, *Vehicle System Dynamics*, 33(Supp), 1999, 374–385, Swets & Zeitlinger, Lisse, The Netherlands.
7. B. Scown, D. Roach, P. Wilson, Freight train driving strategies developed for undulating track through train dynamics research, *Proceedings of the Conference on Railway Engineering*, Adelaide, Australia, 21–23 May 2000, pp. 27.1–27.12.
8. S. Simson, C. Cole, P. Wilson, Evaluation and training of train drivers during normal train operations, *Proceedings of the Conference on Railway Engineering*, Wollongong, Australia, 10–13 November 2002, pp. 329–336.
9. V.K. Garg, R.V. Dukkipati, *Dynamics of Railway Vehicle Systems*, Academic Press, New York, NY, 1984.
10. H.I. Andrews, *Railway Traction: The Principles of Mechanical and Electrical Railway Traction*, Elsevier Science Publishers, New York, NY, 1986.
11. W.W. Hay, *Railroad Engineering* (2nd ed.), John Wiley, New York, NY, 1982, pp. 69–82.
12. V.A. Profillidis, *Railway Engineering* (2nd ed.), Ashgate Publishing, Aldershot, UK, 2000.
13. C. Cole, Improvements to wagon connection modelling for longitudinal train simulation, *Proceedings of the Conference on Railway Engineering*, Rockhampton, Australia, 7–9 September 1998, pp. 187–194.
14. L. Muller, D. Hauptmann, T. Witt, TRAIN—A computer model for the simulation of longitudinal dynamics in trains, *Proceedings of the Conference on Railway Engineering*, Rockhampton, Australia, 7–9 September 1998, pp. 181–186.

15. G.P. Wolf, K.C. Kieres, Innovative engineering concepts for unit train service: The slackless drawbar train and continuous center sill trough train, *Proceedings of the 4th International Heavy Haul Railway Conference*, Brisbane, Australia, 1989, pp. 124–128.
16. G.W. Bartley, S.D. Cavanaugh, The second generation unit train, *Proceedings of the 4th International Heavy Haul Railway Conference*, Brisbane, Australia, 11–15 September 1989, pp. 129–133.
17. C. Cole, M. McClanachan, M. Spiryagin, Y.Q. Sun, Wagon instability in long trains, *Vehicle System Dynamics*, 50(Suppl), 2012, 303–317.
18. Association of American Railroads, *AAR Manual of Standards and Recommended Practices, Section C—Part II, Design, Fabrication, and Construction of Freight Cars*, Section 2.1 Design Data, pp. C-II-9–C-II-34, Washington, DC, 2011.
19. S. Simson, Three axle locomotive bogie steering, simulation of powered curving performance: Passive and active steering bogies, PhD thesis, Central Queensland University, Rockhampton, Queensland, Australia, 2009. See http://hdl.cqu.edu.au/10018/58747.
20. S. Simson, C. Cole, Idealised steering for hauling locomotives, *Rail and Rapid Transit*, 221(2), 2007, 227–236.
21. S. Simson, C. Cole, Simulation of traction and curving for passive steering hauling locomotive, *Rail and Rapid Transit*, 222(2), 2008, 117–127.
22. BS EN 12299:2009, Railway applications—Ride comfort for passengers—Measurement and evaluation, British Standards Institution, London, 2009.
23. UIC Leaflet 513, *Guidelines for Evaluating Passenger Comfort in Relation to Vibration in Railway Vehicles*, UIC Paris, 2003.
24. RDS 7513.3, Railway rolling stock—Interior environment—Part 3: Passenger rolling stock, Draft 2.1, Rail Industry Safety and Standards Board, Canberra, Australia, 2010.
25. AS2670, Evaluation of human exposure to whole body vibration, Standards Australia, Sydney, 1990.
26. AS3860, Fixed guideway people movers, Standards Australia, Sydney, 1991.
27. Association of American Railroads, *AAR Manual of Standards and Recommended Practices, Section S—Casting Details: Specification M-211—Foundry and Product Approval Requirements for the Manufacture of Couplers, Coupler Yokes, Knuckles, Follower Blocks and Coupler Parts*, Washington, DC, 2013.

7 Rail Vehicle–Track Interaction Dynamics

7.1 INTRODUCTION

Understanding rail vehicle dynamics is fundamental to guarantee the safe and cost-effective operations of modern railways. With the increasing demands for safer rail vehicles with higher speeds and higher loads, more innovative methods for rail vehicle dynamics analysis to provide a comprehensive understanding of rail vehicle dynamic performance are required. The application of rail vehicle dynamics analysis encompasses the full range from rail vehicle manufacture (concept development, detailed design, design evolution and risk analysis), to train operations (ride comfort, lateral instabilities, derailment potential) and track infrastructure design and maintenance (maximising the track life and minimising the cost of infrastructure maintenance).

This chapter on rail vehicle dynamics is designed to provide the knowledge required to simulate the dynamic interaction of any rail vehicle with virtually any track. A vehicle and track dynamic system is described by a set of dynamic equilibrium equations for any number of bodies, degrees of freedom and connection elements. Some numerical integration methods are applied to solve the equations. Therefore, the dynamic interactions of rail vehicles and track to predict stability, ride quality, vertical and lateral dynamics and steady state and dynamic curving response can be investigated. The detailed non-linear modelling of wheel–rail interaction, plus secondary and primary suspension responses are included.

7.2 MODELLING OF RAIL VEHICLES

A conventional rail vehicle basically consists of a vehicle car body and two bogies as shown in Figure 7.1.

Two main types of bogies, one without and the other with primary suspensions, are shown in Figure 7.2a and b, respectively. Suspensions are made of coil springs that minimise the impact and enhance the stability of operation of the vehicle.

A simplified design of the bogie known as a *three-piece bogie* is shown in Figure 7.2a. Three-piece bogies consist of a bolster, two sideframes and two wheelsets. Coiled springs are used as the secondary suspensions that connect the bolster with the sideframes. Friction wedges between the sideframes and bolster act as the dampers. Three-piece bogies are the cheapest and most economic to maintain. However, they provide a low level of lateral stability and very poor ride quality. That is why rail vehicles with three-piece bogies have been widely used in freight, coal or mineral

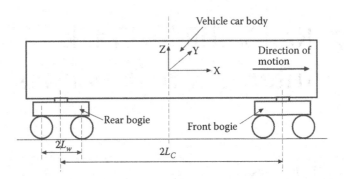

FIGURE 7.1   Schematic side view of a rail vehicle.

train operations. Three-piece bogies, due to higher unsprung mass, generate much larger impact loading. By including coiled springs or rubber between the axles of wheelsets and the sideframes as primary suspensions (Figure 7.2b), the lateral stability of the vehicles and the ride quality are significantly improved at the expense of initial cost. Such bogies, due to lower unsprung mass, generate relatively lower impact loading. That is why such bogies have been widely used in passenger and high-speed train operations.

Other improved designs such as self-steering bogies are also used in heavy haul operations. Passenger cars use much improved designs of bogies suitable for high speed and light load.

In the bogies with primary suspensions, the sideframes provide seating for the secondary suspensions and rest on the two wheelsets through the primary suspensions. In the bogies without primary suspensions, the sideframes rest directly on the wheelsets. It can be seen from Figures 7.1 and 7.2a and b that the vehicle car body rests directly on the centre bowls of the bolsters. This connection allows relative rotation between the bolsters and the vehicle car body about the vertical axis through the centre of each bolster. The bolster is connected to the two sideframes by the secondary suspensions as shown in Figure 7.2a, which allows vertical and lateral displacements in addition to the roll rotation of the bolster together with the vehicle car body relative to the two sideframes. Rail vehicle models can be classified as shown in Figure 7.3.

Figure 7.3 shows that the models for the rail vehicles may be classified into three categories. The first category of the model considers a single wheel with static wheel load moving along the track. The static load represents the mass of all wagon components. The load may either be applied directly on the wheel or through a primary suspension as shown in Figure 7.4.

The second category of the model considers one bogie or 'half car'. In this model, two wheelsets are connected to the sideframes either directly or through primary suspensions. The sideframes support the mass either through the secondary suspension or directly as shown in Figure 7.5.

The third category of the model considers a single wagon that includes all components. In such a model, the wagon car body rests either on the secondary suspensions

(a)

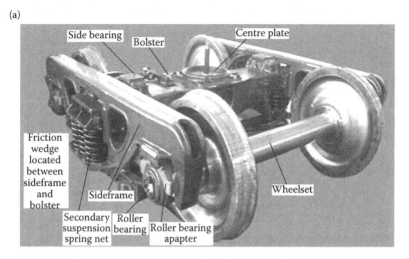

(b)

FIGURE 7.2   Types of bogies. (a) Three-piece freight bogie (no primary suspension), (b) bogie with primary suspension.

through the bolsters or directly on two bogies. The bogies are modelled either with or without primary suspensions.

7.3   MODELLING OF TRACKS

The conventional rail track structure consists of the rail, the fasteners and the pads, the sleepers (ties), the ballast and subballast, and the subgrade. A typical cross section of this type of track structure (used popularly in the heavy haul transport

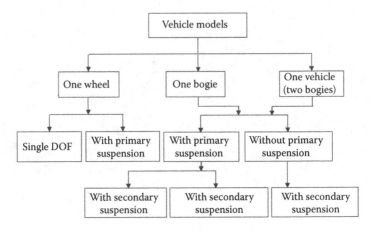

FIGURE 7.3   Model classification for rail vehicles.

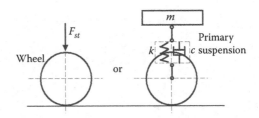

FIGURE 7.4   Single wheel model.

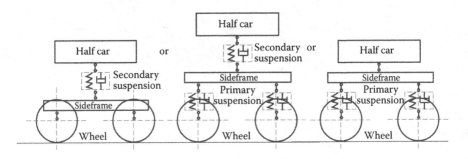

FIGURE 7.5   Bogie models.

network in Australia) is shown in Figure 7.6. The characteristics and the function of each component of the track are described in this section.

Rails support and guide the wheels of the wagon. The widely used rail profile consists of a head, a web and a base and is designated by its weight per unit length (kg/m). The choice of the rail section is made based on the traffic load. Generally, 50, 60 and 68 kg/m rails are used in heavy haul networks. The structural behaviour of

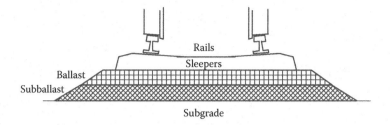

FIGURE 7.6   A typical track structure.

the rail is theoretically modelled as an infinitely long elastic beam resting on elastic supports that are either continuous or discrete. In practice, most modern tracks position the rail on a cant so that the base and the top of the rail slope inwards towards the track centre.

Fasteners connect the rails to the sleepers, whilst *pads* distribute the rail-seat loads evenly to the sleepers. The fasteners ensure that the rail–sleeper connection keeps the track gauge unchanged during traffic operation, in addition to attenuating and dampening the dynamic loads caused by the moving train. Resilient pads are used between the rails and the sleepers. Pads, similar to fasteners, attenuate the transmitted dynamic loads. For assessing system dynamics, the pads and the fasteners are modelled as springs and dampers.

Ballast is the layer of crushed stone on which the sleepers rest. Ballast distributes forces evenly to the subgrade either through the subballast or directly in addition to attenuating a significant part of the dynamic loads. Ballast also resists the shifting of the track superstructure (rail–sleeper assembly) in both the lateral and the longitudinal directions. Note that the ballast usually fully surrounds the sleepers up to their top surface, both between the sleepers and on both ends (not shown in Figure 7.6). The structural properties of ballast are evaluated from both laboratory tests and on-site experiments. For simplicity, ballast is modelled as linear elastic springs and dampers containing mass.

Subballast is a layer immediately below the ballast that contains much finer particles than the ballast layer and performs tasks similar to that of the ballast layer. In most Australian networks, a subballast layer is not present.

The role of the *subgrade* is to withstand traffic load with adequate attenuation. The modulus of elasticity and the carrying capacity of the subgrade are important structural characteristics that classify the subgrade. The ballast and the subballast layers may be considered as a foundation interacting with the subgrade.

The track subsystem models are classified as shown in Figure 7.7.

Figure 7.7 shows that the track modelling may be classified into three types, namely, the continuously supported model, the discretely supported model and the finite element (FE) model. The continuously supported model is based on the beams on elastic foundation (BOEF) theory. The discretely supported model (DSM) allows for the discrete spacing of sleepers. In both approaches, the rail is modelled using either the Euler beam theory or the Timoshenko beam theory. The support for the

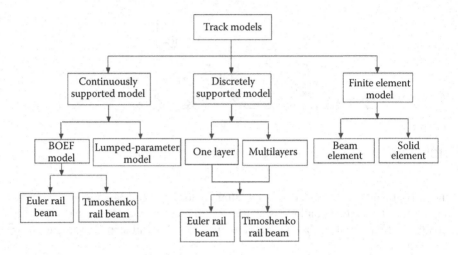

FIGURE 7.7   Model classification for rail track.

rails is modelled either as a single layer or as multiple layers. Multiple layers allow for the inclusion of various track components such as the rails, the pads, the ballast, the subballast and the subgrade.

In the lumped-parameter track model, the rails, the sleepers and the ballast are discretised as lumped masses with lumped stiffnesses and lumped damping coefficients. These lumped properties are evaluated by equating the kinetic energies of the actual and lumped systems.

The FE model is used for more refined stress analysis of track components. Complete FE modelling of the full track system is complicated due to the interface characteristics of the various track components.

7.4   MODELLING OF WHEEL–RAIL CONTACT

At the wheel–rail interface, the contact area and the relationship between the displacement and the normal contact force are determined using Hertz static contact theory. In the tangential direction, the relationship between the creepages and the creep forces is determined using Kalker's creep theory.

The wheelset containing two cone-profiled wheels runs on the rails that are canted inwards at 1 in 20 (or 1 in 40) as shown in Figure 7.8. The gap between

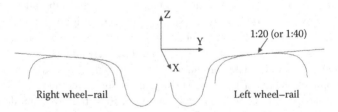

FIGURE 7.8   Wheel–rail interface.

the flange of the wheel and the gauge face of the railhead generally is sufficient to prevent flange contact. Hence, the coned wheelset would have inherent guidance of pure rolling along straight track if it runs on the railhead with no lateral disturbance. However, the guidance of a wheelset on straight track is modified when the wheelset is fitted to a wagon through the suspensions. Furthermore, the pure rolling motion is affected by the action of creep forces tangential to the contact plane between the wheel and the rail surfaces. As the wheelset rolls longitudinally, it also moves laterally and vertically in addition to rotating about the vertical axis. Therefore, the definition of rolling contact between the wheel and the rail becomes fairly complex.

It is well known that the wheel–rail interface creep significantly affects the dynamics of the wagon–track system. The interface creep occurs due to the difference in the velocities of the wheel and the rail at the contact point. The term creepage is used to define the velocity differences in the longitudinal and lateral directions as well as spin creepage due to yaw rotation, with the following expressions for longitudinal, lateral and spin creepage:

$$
\begin{cases}
\xi_x = \dfrac{\begin{array}{c}(\text{Longitudinal velocity of wheel} - \text{longitudinal velocity of rail}) \\ \text{at the point of contact}\end{array}}{\text{Nominal velocity}} \\[2em]
\xi_y = \dfrac{\begin{array}{c}(\text{Lateral velocity of wheel} - \text{lateral velocity of rail}) \\ \text{at the point of contact}\end{array}}{\text{Nominal velocity}} \\[2em]
\xi_{sp} = \dfrac{\begin{array}{c}(\text{Angular velocity of wheel} - \text{angular velocity of rail}) \\ \text{at the point of contact}\end{array}}{\text{Nominal velocity}}
\end{cases}
\tag{7.1}
$$

Note that the longitudinal and lateral creepages are dimensionless, whilst the spin creepage is expressed with the dimension of length^{-1}.

Based on the wheelset and its Cartesian coordinate systems shown in Figure 7.9a and b, the velocities at the contact points on two wheels (assuming small angular displacements ϕ_{wx} about X_w and ϕ_{wz} about Z_w, and the insignificance of their higher-order terms) can be deduced as

$$
\dot{R}_{wi} = (\dot{u}_w - y_{wi}\cos\phi_{wx}\cos\phi_{wz}\dot{\phi}_{wz})\bar{i} + (\dot{v}_w + r_{1i}\cos\phi_{wx}\cos\phi_{wz}\dot{\phi}_{wx})\bar{j} \\
+ (\dot{w}_w + y_{wi}\cos\phi_{wx}\dot{\phi}_{wx})\bar{k}
\tag{7.2}
$$

in which subscript $i = l$ or r represents the left or right wheel, $\dot{u}_w, \dot{v}_w$ and $\dot{w}_w$ are the longitudinal, lateral and vertical velocities respectively of the wheelset at mass centre in the coordinate system (X_w, Y_w, Z_w), $\bar{i}, \bar{j}$ and $\bar{k}$ are the unit vectors corresponding to (X_w, Y_w, Z_w), y_{wi} is the lateral coordinate of the left or right contact point from the wheelset mass centre, and r_{1i} is the rolling radius of the left or right wheel.

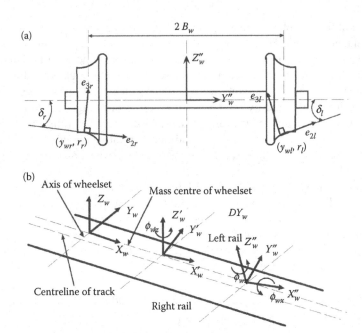

FIGURE 7.9 Wheelset coordinate system. (a) Wheelset, (b) wheelset axes.

Based on the rail and its Cartesian coordinate systems shown in Figure 7.10, the velocities at the contact points on two rails (assuming small angular displacements ϕ_{Rix} about X_{Ri} and the insignificance of its higher-order terms) can be deduced as follows:

$$\dot{R}_{Ri} = \dot{u}_{Ri}\bar{i}_r + (\dot{v}_{Ri} - z_{Ri}\cos\phi_{Rix}\dot{\phi}_{Rix})\bar{j}_r + (\dot{w}_{Ri} + y_{Ri}\cos\phi_{Rix}\dot{\phi}_{Rix})\bar{k}_r \qquad (7.3)$$

in which $\bar{i}_r, \bar{j}_r$ and $\bar{k}_r$ are the unit vectors corresponding to their coordinate systems (X_{Ri}, Y_{Ri}, Z_{Ri}), $\dot{u}_{Ri}, \dot{v}_{Ri}$ and $\dot{w}_{Ri}$ are the longitudinal, lateral and vertical velocities respectively of the rail at mass centre in the coordinate system (X_{Ri}, Y_{Ri}, Z_{Ri}), y_{Ri} and z_{Ri} are the lateral and vertical coordinates respectively of the left or right contact point from the rail neutral centre in (X_{Ri}, Y_{Ri}, Z_{Ri}).

Wheel–rail creep has the potential to cause rail vehicle instability, commonly known as hunting. Rail vehicle instability is usually analysed without due consideration of the velocities of the rail affected by the discrete support of the sleepers (rail track is usually regarded as rigid or continuously supported on an elastic foundation). However, the modelling presented offers a chance to include the velocities of the rail in the lateral and spin directions as required by Equations 7.2 and 7.3. The creepages are used to determine the creep forces and moments using Kalker's linear theory [1]. As Kalker's linear theory best defines the creep forces for very small creepages, the Johnson–Vermeulen approach is used to further modify the creep forces calculated using Kalker's linear theory [2,3]. However, the creep moments cannot be modified due to limitations in the Johnson–Vermeulen theory. The lack of availability of a more 'generalised' theory on this subject matter is apparent here.

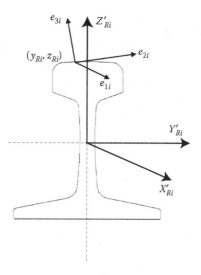

FIGURE 7.10   Rail.

So far, the velocities of contact points on both left and right wheels and rails have been determined with reference to the corresponding coordinate systems. Since the longitudinal dynamics is not considered in the modelling, the longitudinal motion of track is not taken into account and hence $\dot{u}_{Ri}$ in Equation 7.3 is set as zero. By following the definition of creepages for the wheel–rail contact in Equation 7.1, the creepages are determined as follows:

$$
\begin{cases}
\xi_x'' = [\dot{R}_{wi} \cdot e_{1i} - V(r_{1i}/r_0)\cos\phi_{wz}]/V \\[4pt]
\xi_y'' = [(\dot{R}_{wi} - \dot{R}_{Ri}) \cdot e_{2i}]/V \\[4pt]
\xi_{sp}'' = [(\omega_w - \omega_{Ri}) \cdot e_{3i}]/V
\end{cases}
\tag{7.4}
$$

where

$$
\begin{cases}
e_{1i} = \cos\phi_{wz}\,\overline{i} + \sin\phi_{wz}\,\overline{j} \\[4pt]
e_{2i} = -\cos(\delta_i \pm \phi_{wx})\sin\phi_{wz}\,\overline{i} + \cos(\delta_i \pm \phi_{wx})\cos\phi_{wz}\,\overline{j} \pm \sin(\delta_i \pm \phi_{wx})\overline{k} \\[4pt]
e_{3i} = \pm\sin(\delta_i \pm \phi_{wx})\sin_{wz}\overline{i} \mp \sin(\delta_i \pm \phi_{wx})\cos_{wz}\overline{j} + \cos(\delta_i \pm \phi_{wx})\overline{k}
\end{cases}
$$

$$
\begin{cases}
\omega_w = \dot{\phi}_{wx}\overline{i} + (\Omega + \dot{\phi}_{wz}\sin\phi_{wx})(\overline{j} + \dot{\phi}_{wz}\cos\phi_{wx})\overline{k} \\[4pt]
\omega_{Ri} = \phi_{Rix}\overline{i_r} + \phi_{Riy}\overline{j_r} + \phi_{Riz}\overline{k_r}
\end{cases}
$$

in which $\Omega = V/r_0$ is the nominal angular velocity (r_0 is the nominal wheelset rolling radius), the superscript $''$ in Equation 7.4 means that the creepages are defined with respect to the contact plane, and the subscript $i = l$, r represents the left and the right sides respectively, the symbols '$\pm$' and '$\mp$' assume '$+$' and '$-$' when $i = l$, but '$-$' and '$+$' when $i = r$. By simplifying the algebraic expression, neglecting the higher-order terms, assuming small roll angles ϕ_{wx} and yaw angles ϕ_{wz}, and assuming small contact angles, the creepages are expressed as follows:

$$
\begin{cases}
\xi_{ix}'' = (1/V)[V(1 - r_{1i}/r_0) - y_{wi}\dot{\phi}_{wz}] \\
\xi_{iy}'' = (1/V)[(\dot{v}_w + r_{1i}\dot{\phi}_{wx} - V\phi_{wz}) - (\dot{v}_{Ri} - z_{Ri}\dot{\phi}_{Rix})] \\
\xi_{isp}'' = (1/V)[\dot{\phi}_{wz} \mp \Omega\delta_i - \dot{\phi}_{Riz}]
\end{cases}
\tag{7.5}
$$

In Equation 7.5, the symbols '$\pm$' and '$\mp$' assume '$+$' and '$-$' when $i = l$, but '$-$' and '$+$' when $i = r$. In the calculation of the creepages of the left and the right wheel–rail contact around the equilibrium position, Equation 7.5 can be used because the contact angle δ_i is small.

For very small creepages ξ_x'', ξ_y'', and ξ_{sp}'', Kalker's linear creep theory is used to develop the relationships between the creep forces and the creepages. The creep forces and the creep moments on the contact plane (e_{1i} e_{2i}) are obtained as follows:

$$
\begin{cases}
F_{cix}'' = -(f_{33}/V)[V(1 - r_{1i}/r_0) - y_{wi}\dot{\phi}_{wz}] \\
F_{ciy}'' = -(f_{11}/V)[(\dot{v}_w + r_{1i}\dot{\phi}_{wx} - V_{wz}) - (\dot{v}_{Ri} - z_{Ri}\dot{\phi}_{Rix})] - (f_{12}/V)[\dot{\phi}_{wz} \mp \Omega\delta_i - \dot{\phi}_{Riz}] \\
M_{ciz}'' = -(f_{12}/V)[(\dot{v}_w + r_{1i}\dot{\phi}_{wx} - V_{wz}) - (\dot{v}_{Ri} - z_{Ri}\dot{\phi}_{Rix})] - (f_{22}/V)[\dot{\phi}_{wz} \mp \Omega\delta_i - \dot{\phi}_{Riz}]
\end{cases}
\tag{7.6}
$$

in which f_{11}, f_{12}, f_{22} and f_{33} are the creep coefficients.

Kalker's linear theory is limited to the case of small creepages. For large creepages, a simplified approximate model illustrated in [2] is used. In this model, the creep forces are first computed by using Kalker's linear theory (Equation 7.6) and the non-linear effect of adhesion limit is included by computing a resultant force F_{re} as follows:

$$
F_{re} = (F_{cx}''^2 + F_{cy}''^2)^{1/2}
\tag{7.7}
$$

By following the Johnson–Vermeulen approach [2] for creep without spin, the limiting resultant force $\bar{F}_{re}$ is determined by

$$
\bar{F}_{re} =
\begin{cases}
\mu_k F_{WRn}\left[\dfrac{F_{re}}{\mu_k F_{WRn}} - \dfrac{1}{3}\left(\dfrac{F_{re}}{\mu_k F_{WRn}} \right)^2 + \dfrac{1}{27}\left(\dfrac{F_{re}}{\mu_k F_{WRn}} \right)^3 \right] & \text{for} \quad F_{re} \leq 3\mu_k F_{WRn} \\
\mu_k F_{WRn} & \text{for} \quad F_{re} > 3\mu_k F_{WRn}
\end{cases}
\tag{7.8}
$$

in which μ_k is the coefficient of friction at the wheel–rail contact, F_{WRn} is the normal force at the wheel–rail contact.

For small creepages when $F_{re} \leq 3\mu_k F_{WRn}$ and for complete slip when $F_{re} > 3\mu_k F_{WRn}$, the creep forces are given as

$$
\begin{cases}
F_{cx}'' = \dfrac{F_{cx}''}{F_{re}}\bar{F}_{re}, & F_{cy}'' = \dfrac{F_{cy}''}{F_{re}}\bar{F}_{re} & \text{for} \quad F_{re} \leq 3\mu_k F_{WRn} \\[3mm]
F_{cx}'' = \dfrac{\xi_x''}{\tau_{re}}\mu_k F_{WRn}, & F_{cy}'' = \dfrac{\xi_y''}{\tau_{re}}\mu_k F_{WRn} & \text{for} \quad F_{re} > 3\mu_k F_{WRn}
\end{cases}
\tag{7.9}
$$

in which $\tau_{re} = (\xi_x''^2 + \xi_y''^2)^{1/2}$.

After coordinate transformation, the creep forces and the creep moments in equilibrium wheelset and rail coordinate systems are given as

$$
\begin{cases}
F_{cix} = F_{cix}''\cos\phi_{wz} - F_{ciy}''\cos(\delta_i \pm \phi_{wx})\sin\phi_{wz} \\[1mm]
F_{ciy} = F_{cix}''\sin\phi_{wz} - F_{ciy}''\cos(\delta_i \pm \phi_{wx})\cos\phi_{wz} \\[1mm]
F_{ciz} = \pm F_{ciy}''\sin(\delta_i \pm \phi_{wx}) \\[1mm]
M_{cix} = \pm M_{ciz}''\sin(\delta_i \pm \phi_{wx})\sin\phi_{wz} \\[1mm]
M_{ciy} = \mp M_{ciz}''\sin(\delta_i \pm \phi_{wx})\cos\phi_{wz} \\[1mm]
M_{ciy} = M_{ciz}''\cos(\delta_i \pm \phi_{wx})
\end{cases}
\tag{7.10}
$$

The normal contact force between the wheel and the rail is usually determined by using the Hertz static contact theory [4] due to its simplicity [5–10]. In this theory, the contacting bodies are assumed to remain elastic throughout the contact process and the contacting surfaces are assumed smooth and frictionless. In these applications of Hertz static contact theory, it has been assumed that a single point contact occurs on the vertical centreline of the wheel as shown in Figure 7.11. The normal contact force F_{WRni} of the wheel–rail system can be expressed as

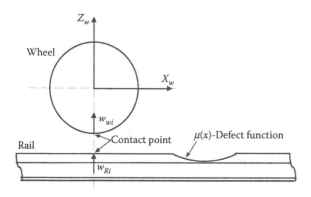

FIGURE 7.11 Wheel–rail single-point contact.

$$F_{WRni} = \begin{cases} C_H[w_{Ri} - w_{wi} - \mu(x)]^{3/2} & \text{if} \quad w_{Ri} - w_{wi} - \mu(x) \geq 0 \\ 0 & \text{if} \quad w_{Ri} - w_{wi} - \mu(x) < 0 \end{cases} \quad (7.11)$$

in which the subscript $i = l, r$ represents the left or the right wheel, C_H is the Hertz contact coefficient in which the subscript H represents the Hertz theory, w_{Ri} is the vertical displacement of the rail at the contact point, w_{wi} is the vertical displacement of the wheel at the contact point and $\mu(x)$ is the function representing the defects of the wheel and/or the rail or the track irregularities.

7.5 EXAMPLE OF A THREE-DIMENSIONAL RAIL WAGON–TRACK SYSTEM DYNAMICS MODEL

The three-dimensional wagon-track system dynamics (WTSD) model includes three subsystems, namely, the wagon subsystem, the track subsystem and the wheel–rail interface subsystem [11–16]. Their dynamic characteristics are briefly described in this section. Detailed formulation of the basic governing equations is similar to a simplified two-dimensional version of this model that has been published [11].

7.5.1 WAGON SUBSYSTEM

The wagon subsystem includes one wagon car body containing two bolsters and two bogies. Each bogie consists of two secondary suspension elements, two sideframes, four primary suspension elements and two wheelsets, as shown in Figure 7.12a,b,c and d.

All wagon components are modelled as rigid bodies with six degrees of freedom (DOFs) (lateral, vertical and longitudinal displacements, and roll, pitch and yaw rotations). The wagon car body, as shown in Figure 7.12a, rests on two bolsters through two centre bowls and four constant-contact side bearings, and is longitudinally connected with two couplers, which are represented as springs. Each centre bowl is modelled with four-point contacts through spring and friction elements along the longitudinal, lateral and vertical directions. The constant-contact side bearing is simplified as spring elements in the vertical direction. The bolster as shown in Figure 7.12b is supported by the suspensions. The sideframe as shown in Figure 7.12c is an intermediate structure that provides seating for the suspensions and connects to the wheelsets shown in Figure 7.12d through steel–steel contacts that are represented as springs with large stiffness. Two kinds of bogie rotations are also taken into account—yaw and lozenge rotations. Non-linear connection characteristics such as vertical lift-off and lateral and longitudinal impacts between sideframe and wheelset, sideframe and bolster, and bolster and wagon car body are fully considered. In Figure 7.12, x represents each DOF, and its subscript indicates the number of DOFs. A total of 66 DOFs are used to describe the movements of all wagon components. For the wagon car body, the centre bowl connection is considered as a four-point lift-off, modelled as a steel–steel contact between car

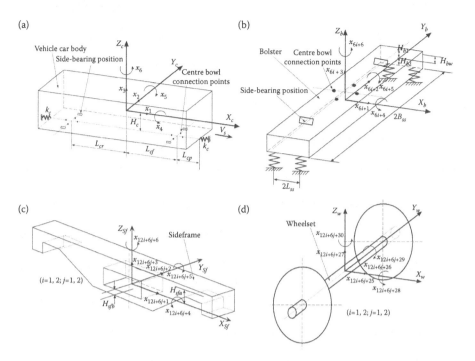

FIGURE 7.12   Wagon model. (a) Car body, (b) bolster, (c) sideframe, (d) wheelset.

body and bolster along X, Y and Z directions. Side bearings are also included as a lift-off connection as shown in Figure 7.12a and b. The constant-damping friction wedges are taken into account in the modelling but are not shown in Figure 7.12b and c. The force and displacement relationships at some connections are described in Figure 7.13.

The forces along the longitudinal X and lateral Y directions at the centre bowl connections between car body and bolsters are calculated according to Figure 7.13a in which Δu_0 and Δv_0 represent the longitudinal and lateral clearances at the centre bowl; F_{0xy} is the friction force; k_{0x} and k_{0y} are the longitudinal and lateral stiffness coefficients of centre bowl contacts. Forces along the vertical Z direction are calculated according to Figure 7.13b in which Δw_0 is the static compression at the vertical contact of the centre bowl; F_{0z} is the preload and k_{0z} is the vertical stiffness coefficient. The total suspension force coming from friction wedges and secondary suspension spring connections between bolster and sideframes are determined based on Figure 7.13c. The suspension spring force and the vertical forces between sideframes and wheelset are calculated according to a relationship similar to that expressed in Figure 7.13b.

The equations of dynamic equilibrium can be written using the multibody mechanics method as

$$M_W \ddot{d}_W + C_W \dot{d}_W + K_W d_W = F_{WT} \tag{7.12}$$

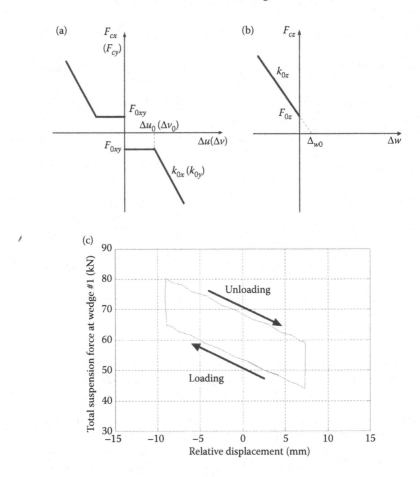

FIGURE 7.13 Force–displacement relationships. (a) Longitudinal and lateral, (b) vertical, (c) total suspension force.

where M_W, C_W and K_W are the mass, damping and stiffness matrices of the wagon subsystem, d_w is the displacement vector of the wagon subsystem and F_{WT} is the interface force vector between the wagon and the track subsystems consisting of the wheel–rail normal contact forces, tangent creep forces and creep moments about the X, Y and Z axes, respectively.

7.5.2 TRACK SUBSYSTEM

The track subsystem model containing four layers is based on the discretely supported distributed parameter approach. The schematic views of the model in the longitudinal and lateral directions are shown in Chapter 5, Figure 5.12a and b, respectively.

In the model, all the track components used in the conventional ballasted heavy haul track structure are assembled in the same sequence as that of the actual structure. The track subsystem comprises two rails, n_S sleepers, $4 \times n_S$ fastener and pad assemblies, $2 \times n_S$ ballast blocks, $2 \times n_S$ subballast blocks and the subgrade.

The lateral and vertical bending and shear deformations of the rail beam are described using Timoshenko beam theory, extended by considering the torque of the rail beam. Thus, 5 DOFs at any point along the longitudinal neutral axis of the rail beam, namely, lateral and vertical displacements and rotations about the lateral (Y) and vertical (Z) axes and torsional rotation about the longitudinal (X) axis, are used in the formulation of the rail beam. For simplicity, the dynamic equilibrium equations of the rail beam have been expanded using a Fourier series in the longitudinal (X) direction by assigning an equal number of terms n_m (also known as the number of deflection modes of the rail beam) for both the linear displacements and the angular rotations. The equations for the vertical deflection and rotation of the rail about the lateral (Y) axis at any point under the action of forces can be expressed as

$$
\begin{cases}
\rho A \dfrac{\partial^2 w_R}{\partial t^2} - GAk\left(\dfrac{\partial^2 w_R}{\partial x^2} - \dfrac{\partial \phi_R}{\partial x} \right) = -\displaystyle\sum_{i=1}^{N_s} F_{RSi}\delta(x - x_i) + \sum_{j=1}^{4} F_{WRnj}\delta(x - x_j) \\[4mm]
\rho I \dfrac{\partial^2 \phi_R}{\partial t^2} - GAk\left(\dfrac{\partial w_R}{\partial x} - \phi_R \right) - EI \dfrac{\partial^2 \phi_R}{\partial x^2} = 0
\end{cases}
\tag{7.13}
$$

where w_R is the vertical deflection of the rail, ϕ_R is the rotation of the rail, ρ is the rail density, A is the area of the rail cross-section, G is the shear modulus of the rail, E is Young's modulus of the rail material, I is the second moment of area of the rail section, k is the Timoshenko shear coefficient, F_{RSi} is the reaction force between the rail and the ith sleeper, F_{WRnj} is the contact force between the jth wheel and the rail, $\delta(x)$ is the Dirac delta function, x_i is the position of the ith sleeper, x_j is the position of the jth wheel and N_s is the number of sleepers considered. The subscript i is used for the sleeper count and j for the wheel count.

The vertical deflection w_R and rotation ϕ_R of the rail are obtained using modal superposition as

$$
\begin{cases}
w_R = \displaystyle\sum_{h=1}^{N_c} N_w(h, x) \cdot W_h(t) \\[4mm]
\phi_R = \displaystyle\sum_{h=1}^{N_c} N_\phi(h, x) \cdot \Phi_h(t)
\end{cases}
\tag{7.14}
$$

where $N_w(h, x)$ and $N_\phi(h, x)$ are the hth mode shape functions of the vertical deflection and rotation respectively of the rail, $W_h(t)$ and $\Phi_h(t)$ are the hth mode time coefficients of the vertical deflection and rotation respectively of the rail, N_c is the number of modes considered and x represents the linear coordinate along the length of the rail beam.

By substituting Equation 7.14 into Equation 7.13, we modify the partial differential Equation 7.13 into the ordinary differential equation shown in Equation 7.15.

This transformation facilitates the application of a numerical method to solve the equations.

$$
\begin{cases}
\dfrac{d^2 W_h}{dt^2} + \dfrac{Gk}{\rho}\left(\dfrac{\Pi h}{L}\right)^2 W_h - \sqrt{\dfrac{A}{I}}\dfrac{Gk}{\rho}\left(\dfrac{\Pi h}{L}\right)\Phi_h = -\sum_{i=1}^{N_s} F_{RSi}N_w(h,x_i) + \sum_{j=1}^{4} F_{WRnj}N_w(h,x_j) \\[3mm]
\dfrac{d^2 \Phi_h}{dt^2} + \left(\dfrac{GAk}{\rho I} + \dfrac{E}{\rho}\left(\dfrac{\Pi h}{L}\right)^2\right)\Phi_h - \sqrt{\dfrac{A}{I}}\dfrac{Gk}{\rho}\left(\dfrac{\Pi h}{L}\right)W_h = 0
\end{cases}
$$

$$(h = 1,2,...,N_c) \qquad (7.15)$$

in which L is the length of the rail considered, and the reaction force between the rail and the ith sleeper, F_{RSi}, is expressed as

$$
F_{RSi} = (C_{pi} + C_{fi})\sum_{m=1}^{N_c} N_w(m,x)\cdot \dot{W}_m + (K_{pi} + K_{fi})\sum_{m=1}^{N_c} N_w(m,x)\cdot W_m
$$
$$
- (C_{pi} + C_{fi})\cdot \dot{w}_{si} - (K_{pi} + K_{fi})\cdot w_{si} \qquad (7.16)
$$

where w_{si} is the vertical displacement of the ith sleeper. C_{pi}, K_{pi} and C_{fi}, K_{fi} are the damping and stiffness coefficients of the ith pad and the ith fastener, respectively.

In Equation 7.15, the contact force F_{WRnj} between the jth wheel and the rail is determined by non-linear Hertz contact theory and is given in Equation 7.11.

The sleepers are considered as deformable short beams resting on an elastic foundation and represented by their mass and viscoelastic properties at the rail seat location, and having three DOFs per sleeper, namely, the lateral and vertical displacements and the rotation about the longitudinal (X) direction. The ballast and the subballast layers are represented by their mass and viscoelastic properties, with only one vertical displacement DOF per block. The properties of the ballast and subballast have been determined from a pyramidal model as described below.

The ballast ensures damping of the vibrations and distributes the load evenly to the subgrade. The subballast protects the top surface of the subgrade from penetration by the ballast stone particles in addition to further distributing the load. A ballast pyramid model based on the theory of elasticity was developed [17]. The ballast–subballast pyramid model assumes that the loading and pressure distribution is uniform throughout the depth. The model is divided into the upper and lower sections, which reflects the actual transmission of the loading. The vibration of the ballast was defined as a single block [18] based on the observation that the accelerations of the individual particles in both upper and lower surfaces of the ballast block do not vary significantly even though such a conclusion is not universal. The oscillating mass of each ballast block is calculated by multiplying the volume of the ballast block by the ballast density. According to [17], the stiffness of each ballast block K_{bl} is

$$K_{bl} = \frac{2\tan\theta_b(L_s - B_s)E_b}{\ln\left[(L_s(2\tan\theta_b H_b + B_s))/(B_s(2\tan\theta_b H_b + L_s))\right]} \qquad (7.17)$$

in which L_s and B_s are the effective length and width of the support area of the rail seat, E_b is the modulus of elasticity of the ballast (in N/m^2), θ_b is the internal friction angle of ballast (20° is chosen for ballast as suggested in [17]), and H_b is the height of the ballast.

Similarly, the stiffness of each subballast block K_{sb} is

$$K_{sb} = \frac{2\tan\theta_{sb}(L_s - B_s)E_{sb}}{\ln\left[\begin{array}{l}((2\tan\theta_b H_b + L_s)(2\tan\theta_{sb}H_{sb} + 2\tan\theta_b H_b + B_s)) \\ /((2\tan\theta_b H_b + B_s)(2\tan\theta_{sb}H_{sb} + 2\tan\theta_b H_b + L_s))\end{array}\right]} \qquad (7.18)$$

in which E_{sb} is the modulus of elasticity of the subballast (in N/m^2), θ_{sb} is the internal friction angle of subballast (35° is chosen for subballast) and H_{sb} is the height of the subballast.

The damping coefficients of the ballast and the subballast are determined as 40% of their critical damping coefficients. This damping ratio (40%) is considered realistic for earth structures and it is found that these values are within the range given by [19] (e.g. the post-tamping and pre-tamping tracks were 30 and 82 kNs/m, respectively).

The oscillating masses of each ballast block M_{bl} and subballast block M_{sb} are

$$M_{bl} = \rho_b\left[L_sB_s + H_b\tan\theta_b(L_s + B_s) + \frac{4}{3}H_b^2\tan^2\theta_b\right]$$

$$M_{sb} = \rho_{sb}\left[(L_s + 2\tan\theta_b)(B_s + 2\tan\theta_b) + H_{sb}\tan\theta_{sb}(L_s + B_s + 4\tan\theta_b) + \frac{4}{3}H_{sb}^2\tan^2\theta_{sb}\right]$$

$$(7.19)$$

The subgrade stiffness K_{sg} is

$$K_{sg} = E_{sg}(2\tan\theta_{sb}\cdot H_{sb} + 2\tan\theta_b\cdot H_b + L_s)(2\tan\theta_{sb}\cdot H_{sb} + 2\tan\theta_b\cdot H_b + B_s)$$

$$(7.20)$$

in which E_{sg} is the modulus of the subgrade expressed in N/m^3.

In the longitudinal direction, the continuity of the ballast and the subballast are ensured by including viscoelastic elements (without mass) connecting the blocks of ballast and subballast in their respective layers. The coefficients of these longitudinal springs and dampers were calculated by multiplying the respective vertical stiffness and damping coefficients by a factor of 0.3. This factor is not sensitive to the dynamic responses on the interface between the wagon and the track. The ballast and the subballast layers are connected in the vertical and the horizontal (both longitudinal and lateral) directions to depict the continuity of these layers. The subgrade (formation) is represented by its viscoelastic properties without mass. The equations of dynamic equilibrium of the sleepers, the ballast, the subballast and the subgrade are assembled using multibody mechanics methods. Finally, the governing equations of dynamic equilibrium for the track (rail and all other track components) are expressed in the following matrix form:

$$M_T \ddot{d}_T + C_T \dot{d}_T + K_T d_T = \tilde{F}_{WT} \tag{7.21}$$

in which M_T, C_T and K_T [each of size $(10n_m + 7n_s) \times (10n_m + 7n_s)$] are the mass, damping and stiffness matrices of the track subsystem, respectively. The vector d_T contains the displacement of the track subsystem which includes the modal and physical displacements and $\tilde{F}_{WT}$ is the combined interface force vector between the wagon and the track subsystems.

Selection of n_s and n_m depends on the required length of travel of the wagon to be simulated. For a travelling distance of 40 sleeper spacings, it was found that at least a track length of 120 sleeper spacings should be considered for the results not to be affected by the boundary conditions of the rail beams and the initial conditions used in the dynamic simulation.

7.6 NUMERICAL INTEGRATION METHODS

The equations of motion of the wagon and track system could be solved either in the frequency domain or in the time domain. Solving the equations using the frequency-domain method is usually simpler and more efficient than under the time domain. However, all calculations using frequency-domain methods tacitly assume that the whole wagon and track system is completely linear. There are actually a lot of non-linearities in the rail track and wagon system such as the non-linear Hertz contact and the creepages. Therefore, a numerical integration method in the time domain is used for the solution of Equations 7.12 and 7.21.

The direct numerical integration methods in the time domain can be adopted because, in these methods, the equations of motion are integrated successively using a step-by-step numerical procedure with no transformation of the equations of motion being necessary prior to the integration.

There are two basic approaches used in the direct numerical integration methods, namely, the explicit and the implicit algorithms. In an explicit formulation, the response quantities are expressed in terms of previously determined displacement, velocity and acceleration. Some explicit schemes such as the fourth-order Runge–Kutta method and the central difference method are often used. In an implicit formulation, the equations of motion are directly solved for the displacements. The implicit schemes include Houbolt, Wilson-θ and Newmark-β methods.

An example of a wheelset running on rigid rails is contained in [2] to illustrate the performance of seven integration methods—central difference predictor, two-cycle iteration with trapezoidal rule, fourth-order Range–Kutta methods (explicit schemes) and Houbolt, Wilson-θ, Newmark-β and Park's Stiffly stable methods (implicit schemes). The results were calculated under the condition of wheelset speed $V = 96.54$ km/h and the time step $= 0.001$ s. From the results of the lateral displacement and the yaw rotation of the wheelset, it was concluded that the Newmark-β and Park's Stiffly Stable methods were suitable for the wagon–track dynamics problem. The Newmark-β method has also been successfully used to solve train–wagon–track dynamics [18,20,21].

The numerical integration methods in the time domain available in commercial rail system dynamics software include a Runge–Kutta–Bettis code and two multistep codes [22]. The fourth-order Runge–Kutta method with adaptive time-step size control was used to solve dynamic wheel–rail and track interactions [6].

For the solutions of the dynamic equations of equilibrium of the wagon and track system presented in this section, two methods—the improved fourth order Runge–Kutta and the modified Newmark-β were selected. This is because the Runge–Kutta method adopts an explicit scheme and the modified Newmark-β method adopts an implicit scheme. The efficiency of these two methods is illustrated below using an example. It should be emphasised that the use of a particular method is largely affected by the nature of the problem and is often dictated by the desired solution accuracy. The selection of the appropriate integration time step is often based on previous experience, but should be evaluated by testing solvers at different time steps. The largest time step that does not result in numerical instability can be chosen.

7.6.1 IMPROVED FOURTH-ORDER RUNGE–KUTTA METHOD

It is well known that the fourth-order Runge–Kutta method has the truncation error $O(h^4)$ (h is the step size) and requires four evaluations of function per step. An improved version of this method, known as the Runge–Kutta–Fehlberg method, has been used in numerical analysis [23,24]. Although this method requires six evaluations of function per step, it has a much better truncation error $O(h^5)$ and can adopt any time step size depending on the value of estimated error. In the Runge–Kutta–Fehlberg scheme, the system equations (Equations 7.12 and 7.21) are converted into the following first-order equations:

$$\begin{cases} \{\dot{s}_W\} = [A_W]\{s_W\} + \{q_W\} = f(s_W,t) \\ \{\dot{s}_T\} = [A_T]\{s_T\} + \{q_T\} = f(s_T,t) \end{cases} \tag{7.22}$$

in which $[A_W]$ and $[A_T]$ are matrices of wagon and track subsystems, respectively, and are given by

$$\begin{cases} [A_W] = \begin{bmatrix} [0] & [I] \\ -[M_W]^{-1}[K_W] & -[M_W]^{-1}[C_W] \end{bmatrix} \\ [A_T] = \begin{bmatrix} [0] & [I] \\ -[M_T]^{-1}[K_T] & -[M_T]^{-1}[C_T] \end{bmatrix} \end{cases} \tag{7.23}$$

in which [0] is the null matrix and [I] is the unit matrix, and

$$\{s_W\} = \begin{Bmatrix} \{d_W\} \\ \{\dot{d}_W\} \end{Bmatrix}, \quad \{s_T\} = \begin{Bmatrix} \{d_T\} \\ \{\dot{d}_T\} \end{Bmatrix}, \quad \{q_W\} = \begin{Bmatrix} \{0\} \\ [M_W]^{-1}[K_W] \end{Bmatrix}, \quad \{q_T\} = \begin{Bmatrix} \{0\} \\ [M_T]^{-1}[K_T] \end{Bmatrix}$$

The algorithm of the Runge–Kutta–Fehlberg method is based on the following formulas:

$$\begin{cases} \{s_{W,m+1}\} = \{s_{W,m}\} + \left(\dfrac{16}{135} k_{W1} + \dfrac{6656}{12,825} k_{W3} + \dfrac{28,561}{56,430} k_{W4} - \dfrac{9}{50} k_{W5} + \dfrac{2}{55} k_{W6} \right) \\[2ex] \{s_{T,m+1}\} = \{s_{T,m}\} + \left(\dfrac{16}{135} k_{T1} + \dfrac{6656}{12,825} k_{T3} + \dfrac{28,561}{56,430} k_{T4} - \dfrac{9}{50} k_{T5} + \dfrac{2}{55} k_{T6} \right) \end{cases}$$

$$(7.24)$$

where k_{W1}, k_{W3}, k_{W4}, k_{W5}, and k_{W6}, and k_{T1}, k_{T3}, k_{T4}, k_{T5}, and k_{T6} are the approximate derivative values computed in the interval $m \cdot h \le t_m \le (m + 1) \cdot h$ (h is time step), and are determined as follows:

$$\begin{cases} k_{W/T1} = h \cdot f(s_{W/T,m}, t_m) \\[1ex] k_{W/T2} = h \cdot f(s_{W/T,m} + k_{W/T1}/4, t_m + h/4) \\[1ex] k_{W/T3} = h \cdot f(s_{W/T,m} + 3k_{W/T1}/32 + 9k_{W/T2}/32, t_m + 3h/8) \\[1ex] k_{W/T4} = h \cdot f(s_{W/T,m} + (1932k_{W/T1} - 7200k_{W/T2} + 7296k_{W/T3})/2197, t_m + 12h/13) \\[1ex] k_{W/T5} = h \cdot f\left(s_{W/T,m} + \left(\dfrac{439}{216} k_{W/T1} - 8k_{W/T2} + \dfrac{3680}{513} k_{W/T3} - \dfrac{845}{4104} k_{W/T4} \right), t_m + h \right) \\[1ex] k_{W/T5} = h \cdot f\left(s_{W/T,m} + \left(-\dfrac{8}{27} k_{W/T1} + 2k_{W/T2} - \dfrac{3544}{2565} k_{W/T3} + \dfrac{1859}{4104} k_{W/T4} \right. \right. \\[2ex] \qquad \left. \left. - \dfrac{11}{40} k_{W/T5} \right), t_m + \dfrac{h}{2} \right) \end{cases}$$

$$(7.25)$$

in which the subscript W/T represents the wagon or the track subsystem, respectively.

7.6.2 Modified Newmark-β Method

The modified Newmark-β method, involving the introduction of the predictor–corrector integration scheme adopted in [18] and [25], is considered here. The algorithm of the modified Newmark-β scheme is provided in this section. The method involves:

- The explicit difference formulas developed in [18] and [25] based on the Newmark-β scheme are firstly used to predict the displacements and the velocities;
- Then Equations 7.12 and 7.21 are used to calculate the predicted values of accelerations;
- The Newmark-β implicit scheme is used to correct the displacements and the velocities;
- Then Equations 7.12 and 7.21 are again used to calculate the corrected values of accelerations.

1. Predict

$$
\begin{cases}
\{d_{W/T}\}_{p,m+1} = \{d_{W/T}\}_m + \{\dot{d}_{W/T}\}_m h + \left(\frac{1}{2} + \psi\right)\{\ddot{d}_{W/T}\}_m h^2 - \psi\{\ddot{d}_{W/T}\}_m h^2 \\
\{\dot{d}_{W/T}\}_{p,m+1} = \{\dot{d}_{W/T}\}_m + (1+\phi)\{\ddot{d}_{W/T}\}_m h - \phi\{\ddot{d}_{W/T}\}_{m-1} h
\end{cases}
\tag{7.26}
$$

2. Calculate

$$
\begin{aligned}
\{\ddot{d}_{W/T}\}_{p,m+1} = [M_W]_{m+1}^{-1}(\{F_{W/T}\}_{m+1} &- [C_{W/T}]_{m+1}\{\dot{d}_{W/T}\}_{p,m+1} \\
&- [K_{W/T}]_{m+1}\{d_{W/T}\}_{p,m+1})
\end{aligned}
\tag{7.27}
$$

3. Correct

$$
\begin{cases}
\{d_{W/T}\}_{m+1} = \{d_{W/T}\}_m + \{\dot{d}_{W/T}\}_m h + \left(\frac{1}{2} - \beta\right)\{\ddot{d}_{W/T}\}_m h^2 - \beta\{\ddot{d}_{W/T}\}_{p,m+1} h^2 \\
\{\dot{d}_{W/T}\}_{m+1} = \{\dot{d}_{W/T}\}_m + (1-\gamma)\{\ddot{d}_{W/T}\}_m h + \gamma\{\ddot{d}_{W/T}\}_{p,m+1} h
\end{cases}
\tag{7.28}
$$

4. Calculate

$$
\begin{aligned}
\{\ddot{d}_{W/T}\}_{m+1} = [M_W]_{m+1}^{-1}(\{F_{W/T}\}_{m+1} &- [C_{W/T}]_{m+1}\{\dot{d}_{W/T}\}_{m+1} \\
&- [K_{W/T}]_{m+1}\{d_{W/T}\}_{m+1})
\end{aligned}
\tag{7.29}
$$

where h is the time step, $(m+1)$, m and $(m-1)$ are the subscripts to denote the integration time $(m+1)h$, mh and $(m-1)h$, respectively, ψ and ϕ, and β and γ are the free parameters that control the stability and the numerical dissipation of the algorithm.

In the first step, ψ and ϕ, and β and γ are set to zero, and subsequently, ψ and ϕ are both assigned a value of 0.5 while β and γ are both set as 1/6.

7.6.3 Comparison of the Two Solution Schemes

The dynamic responses calculated using the two numerical methods (the Runge–Kutta–Fehlberg and modified Newmark-β) are compared. A vertical 2D single wheel/single layer track model as shown in Figure 7.14 is used for this purpose. In this model, a wheel carrying a static load rolls on a rail supported by the track modelled with springs and dampers arranged at intervals equal to the sleeper spacing. The model data are listed in Table 7.1.

The stiffness and damping coefficients of the static equivalent spring and damper of the track are determined using the following formula:

$$
\begin{cases}
\dfrac{1}{K_e} = \dfrac{1}{K_p} + \dfrac{1}{K_b} + \dfrac{1}{K_{sb}} + \dfrac{1}{K_{sg}} \\[2ex]
\dfrac{1}{C_e} = \dfrac{1}{C_p} + \dfrac{1}{C_b} + \dfrac{1}{C_{sb}} + \dfrac{1}{C_{sg}}
\end{cases}
\tag{7.30}
$$

in which K_p and C_p are the stiffness and damping coefficients of the pad listed in Table 7.1, respectively. K_b, K_{sb}, K_{sg} and C_b, C_{sb}, C_{sg} are the stiffness and damping coefficients of ballast, subballast and subgrade calculated using the ballast–subballast pyramid model according to the parameters listed in Table 7.1, respectively. The 2D single wheel/single layer track model results in 201 equations of motion as described below:

* 1 DOF is used for one wheel;
* The number of modes for the rail Timoshenko beams is 100;
* Total DOFs are $1 + 100 \times 2 = 201$.

The rail vertical displacement at the contact point and the wheel–rail contact force factor presented in Figure 7.15a and b, respectively, are obtained with a time step of 1×10^{-5} s using the modified Newmark-β method. The same information calculated using the Runge–Kutta–Fehlberg method is presented in Figure 7.15c and d. It can be seen that both methods predict stable solutions. The difference between the results is very small. The vertical displacement at the point of contact of the rail predicted by the modified Newmark-β method and the Runge–Kutta–Fehlberg method are 1.42719 and 1.42738 mm, respectively (the difference is 0.01%). The contact force factor predicted by both methods is 1.0. Although the results from both

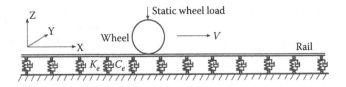

FIGURE 7.14   A 2D single wheel/single layer track model.

TABLE 7.1
Model Parameters for Examining Solution Techniques

Notation	Parameter	Value
Wagon Subsystem		
F_S	Static wheel load	99800 N
m_w	Wheelset mass	1120 kg
I_{wx}, I_{wz}	Mass moment of inertia of wheelset about X, Z axis	420.1 kg·m^2
V	Wheel speed	80 km/h
r_1	Wheel radius	0.425 m
Track Subsystem		
m_r	Rail mass per meter	60 kg/m
A_R	Rail cross-section area	7.77×10^{-3} m^2
E	Elastic modulus of rail	2.07×10^{11} N/m^2
G	Shear modulus of rail	8.1×10^{10} N/m^2
I_{Ry}	Rail second moment of area about Y axis	2.94×10^{-5} m^4
I_{Rz}	Rail second moment of area about Z axis	4.9×10^{-6} m^4
k_R	Timoshenko shear coefficient	0.34
r_R	Rail profile radius on top	0.30 m
n_m	Number of rail Timoshenko modes	100
n_s	Number of sleepers	100
C_{pz}	Pad damping about Z axis	65 kN·s/m
K_{pz}	Pad stiffness about Z axis	450 MN/m
C_{py}	Pad damping about Y axis	26 kN·s/m
K_{py}	Pad stiffness about Y axis	180 MN/m
L_s	Sleeper spacing	0.685 m
L_b	Effective length of support area of the sleeper at a rail seat	1.075 m
B_b	Effective width of support area of the sleeper at a rail seat	0.24 m
H_b	Height of ballast	0.30 m
H_{sb}	Height of subballast	0.15 m
E_b	Elastic modulus of ballast	130×10^6 N/m^2
E_{sb}	Elastic modulus of subballast	200×10^6 N/m^2
E_s	Subgrade modulus	50×10^6 N/m^3
ρ_b	Density of ballast	2600 kg/m^3
ρ_{sb}	Density of subballast	2600 kg/m^3
Interface Subsystem		
C_H	Hertz spring constant	0.87×10^{11} N/m$^{3/2}$

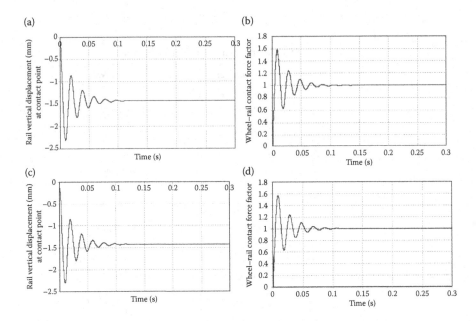

FIGURE 7.15 (a) and (b) Predicted by modified Newmark-β method, and (c) and (d) by the Runge–Kutta–Fehlberg method, both methods at a time step size of 1×10^{-5} s. (a) Rail vertical displacement, (b) wheel–rail contact force factor, (c) rail vertical displacement, (d) wheel–rail contact force factor.

methods are very close to each other, the time consumed by these two methods is different. The modified Newmark-β method took 9 s on a Pentium-4 1.6 GHz computer whilst the Runge–Kutta–Fehlberg method took 14 s on the same platform. This difference is understandable because the Runge–Kutta–Fehlberg method requires six evaluations of $f(y_{W/T}, t)$ per step whilst the modified Newmark-β method requires only two evaluations per step.

The dynamic responses calculated with the time step of 1.5×10^{-5} s using the modified Newmark-β method (Figure 7.16a and b) are still stable, whilst those obtained using the Runge–Kutta–Fehlberg method (Figure 7.16c and d) with the same time step exhibit instability. The time consumed for these two methods are 5 and 9 s, respectively. As the Runge–Kutta–Fehlberg method exhibits instability, it was not used further to investigate the effect of time step. With further increase in time step, the modified Newmark-β method starts to exhibit some sign of instability. It can be seen that, with the time step of 3.5×10^{-5} s, although the rail displacement at the contact point is stable, the wheel–rail contact force is unstable (Figure 7.17a and b). However, with a time step of 5.5×10^{-5} s, both the rail displacement at the contact point and the wheel–rail contact force become unstable.

It could therefore be concluded that the Runge–Kutta–Fehlberg method requires very small time steps to obtain stable solutions and consumes relatively more time for the calculation. Therefore, the modified Newmark-β method has been adopted as the standard solution technique.

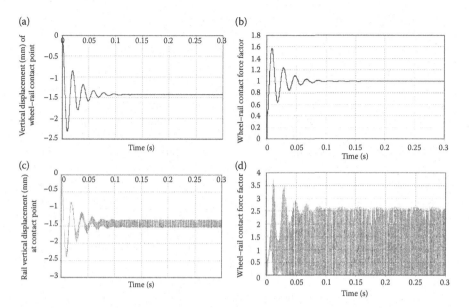

FIGURE 7.16   (a) and (b) Predicted by modified Newmark-β method, and (c) and (d) by the Runge–Kutta–Fehlberg method, both methods at time step size of 1.5×10^{-5} s. (a) Rail vertical displacement, (b) wheel–rail contact force factor, (c) rail vertical displacement, (d) wheel–rail contact force factor.

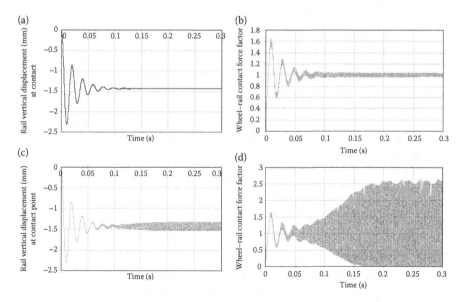

FIGURE 7.17   Predicted by modified Newmark-β method, (a) and (b) at time step size of 3.5×10^{-5} s, and (c) and (d) at time step size of 5.5×10^{-5} s. (a) Rail vertical displacement, (b) wheel–rail contact force factor, (c) rail vertical displacement, (d) wheel–rail contact force factor.

7.7   VEHICLE DYNAMIC PERFORMANCES

7.7.1   DERAILMENTS

A railway vehicle derailment is an accident on a track whereby a train leaves the rails; this can result in significant casualties and property loss. There are many potential causes for derailments including broken rails, severe track geometry irregularities, switch and crossing faults, excessive speed, wheel, bearing or axle failures, bogie failures, collisions with obstructions on the track, poor driving techniques, wagon rollover and so on. The modes for wheels running off rails can be classified into four major categories, these being wheels climbing over the rail, rail gauge widening or rail rollover that both cause the wheels to fall between the rails and track panel lateral shift.

Rail vehicle derailments, excluding derailments due to component failures or overspeed, have been an important issue for dynamics researchers to work on. The rail vehicle derailment phenomena due to wagon dynamics were classified into three types in [26]—climbing-up, slipping-up and jumping-up. Several indices for evaluating safety against derailment were presented as follows:

- Loss of static vertical wheel load (limited to less than 60% of the average static wheel load);
- Loss of dynamic vertical wheel load (limited to less than 80% of the average dynamic wheel load). Wheel unloading is also given a time needed. Events must be either >50 ms (typical freight) or >2 m of track (high-speed train);
- The derailment quotient (L/V) must not exceed 0.8 (L represents the lateral force and V the vertical force during the wheel flange contacting the gauge face of the rail).

Several other safety criteria of incipient derailment which must not be exceeded were also provided by [26–28]. These included varying the maximum allowable derailment quotient L/V according to the wheel flange angle (e.g. L/V of 1.0 for a wheel flange angle of 70° on a Shinkansen high-speed passenger train), and a proviso that, if the time duration for which the value of L/V continued to exceed the limit value was equal to or less than 15 ms, the vehicle would be safe against derailment.

In this section, the derailment criteria due to wheel flange climb as defined by the Nadal formula are discussed. Several safety criteria which are proposed as guidelines for railway operational safety issues based on the Nadal formula are presented, and a derailment simulation replicating the collision of two passenger trains using Gensys software is detailed.

7.7.1.1   Nadal Formula

To estimate the vehicle safety, one can analyse the possibility of derailment. Various formulas exist as a guide for the derailment process, which give the safe range of the ratio between lateral and vertical forces for a particular wheel–rail combination. This ratio, usually called the 'derailment ratio', is denoted as L/V, where L and V are, respectively, the lateral and vertical forces at the flange contact. The derailment ratio L/V is used as a measure of the running safety of railway vehicles. Several theories have been developed to establish the permitted L/V ratio. One of the most widely used is the Nadal formula. This formula takes into account the influence of the wheel flange angle, the wheel–rail friction coefficient and the wheel–rail forces on the possibility of wheel climb derailment. This principle is expressed in the Nadal formula:

$$\frac{L}{V} = \frac{\tan\delta - \mu}{1 + \mu\tan\delta} \tag{7.31}$$

where L = wheel–rail lateral force, V = wheel–rail vertical force, L/V = ratio of wheel–rail forces, δ is the angle between wheel flange contact and the horizontal plane and μ is the friction coefficient.

The explanation of Equation 7.31 is that, at the condition of the maximum wheel–rail contact angle, the minimum value of L/V may likely cause a derailment due to flange climbing.

The limiting L/V ratios for various combinations of friction coefficient and contact angles are shown in Figure 7.18. If the L/V ratio exceeds the limiting value for particular combinations of the friction coefficient and the contact angle, then derailment is likely to occur.

The theory of Nadal is used to establish safe limits for the L/V derailment ratio. The derailment criteria are typically formulated (e.g. in Australian Standard AS7509 [50]) as:

• The maximum individual wheel L/V ratios sustained for 2 m should not exceed 1.0 for vehicles with a soft lateral suspension;
• The maximum individual wheel L/V ratio sustained for 50 ms should not exceed 1.0;
• The maximum sum L/V axle (sum of the absolute values of the individual wheel L/V of both wheels on an axle at the same time) sustained for 50 ms should not exceed 1.5.

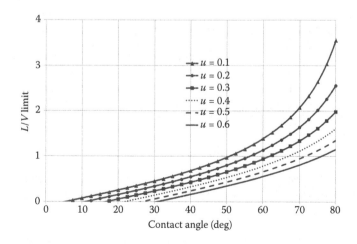

FIGURE 7.18 Limiting L/V ratios.

7.7.1.2 Derailment Simulation Due to Train Collision

The following derailment simulation is detailed as an example of the general analysis technique [29,30]. The scenario is that a 5-car passenger train is travelling at a speed of $V_1 = 80$ km/h and approaching a 2-car passenger train that is stationary in front of it. In fact, Gensys cannot allow the speed to be zero, so we set $V_2 = 0.1$ km/h. A very small deviation of the 5-car passenger train's centreline from that of the 2-car passenger train is set, for example, 0.005 m. Figure 7.19 shows the crash simulation visualisation.

From the simulation results, the first car of the 2-car passenger car train is one that is derailed. Figure 7.20 shows the wheel–rail contacts of the first and second wheelsets in the trailing bogie of that first passenger car (the first wheelset is shown above the second wheelset in each simulation output; left and right wheels are as viewed in these outputs). The collision occurs at time t = 0.1 s. At time t = 0.35 s, the left wheel of the second wheelset commences flange contact; by t = 0.36 s, it has climbed onto the head of the rail. The first wheelset commences lifting up at t = 0.37 s. The left wheel of the first wheelset has started to climb onto the head of the rail at t = 0.47 s. By time t = 0.69 s, both wheelsets have completely derailed.

7.7.2 Rail Ride Comfort

Rail ride quality is a person's reaction to a set of physical conditions in a rail vehicle environment, such as dynamic, ambient and spatial conditions [31,32]. Concerning the dynamic variables only, they deal with the rail vehicle motions, usually measured as accelerations and changes (jerk) in accelerations in all three directions (longitudinal, lateral and vertical), angular motions about these directions (roll, pitch and yaw) and sudden motions, such as shocks and jolts. Rail ride comfort is usually understood to refer to the technical evaluation of dynamic quantities (motions of the rail vehicle).

There are many ride quality standards available to assess the passenger comfort on trains. These standards indicate how major railway organisations and authorities around the world measure and assess the rail vehicle ride quality and comfort, and state what measurements and analysis methods should be used to accurately calculate the ride comfort and relate it to rail vehicle dynamics and longitudinal train dynamics. The following are some major standards:

- BS EN 12299:2009 Railway applications. Ride comfort for passengers. Measurement and evaluation [33,34];
- ISO 2631 Mechanical vibration and shock—Evaluation of human exposure to whole-body vibration [35,36];
- RDS 7513.3, RISSB, Australia [37];
- RTRI Indices, Japan [38,39];
- Sperling Ride Index, Germany [40].

In the standard BS EN 12299:2009 [33] stipulated by The European Committee for Standardization (CEN), for example, there are the following indices for assessing rail ride comfort:

- The mean comfort standard method N_{MV}, which is used to quantify seated passenger comfort during a continuous 5-min run. The accelerations are

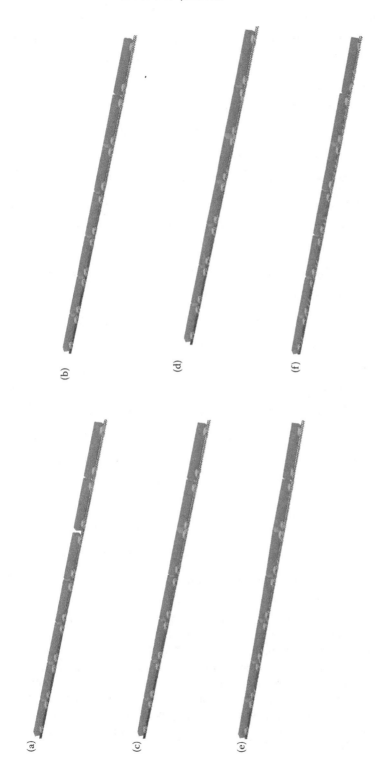

FIGURE 7.19 Derailment simulation visualisation. (a) Two train collision, before collision, (b) collision starts, (c) crumpling between first car of rear train and trailing car of front train, (d) crumpling among cars, (e) lifting up between first car of rear train and trailing car of front train, (f) first car of front train derailed.

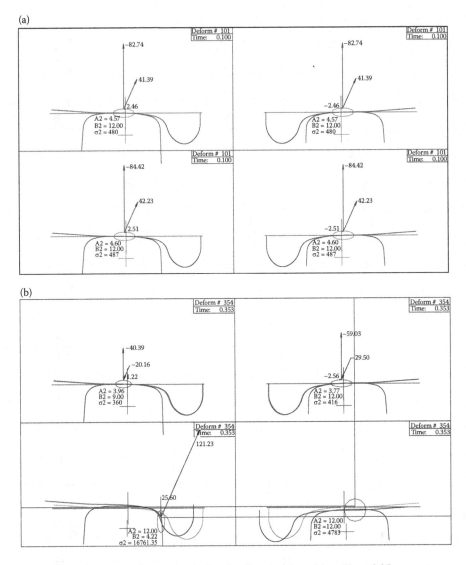

FIGURE 7.20   Wheel-rail contacts during derailment. (a) $t = 0.1$ s, (b) $t = 0.35$ s.

measured in the longitudinal (x), lateral (y) and vertical (z) directions, with frequency weighting in the frequency range from 0.4 to 100 Hz.

- Mean comfort complete methods N_{VA} and N_{VD}. The N_{VA} is used to quantify comfort during a continuous 5-min run for seated passengers, based on accelerometer measurements both on the floor (vertical direction) and in the interfaces between a seated passenger and the seat pan (lateral and vertical directions) and seat back (longitudinal direction). The N_{VD} method is for standing passengers, based on accelerations measured on the floor only.

(c)

(d)

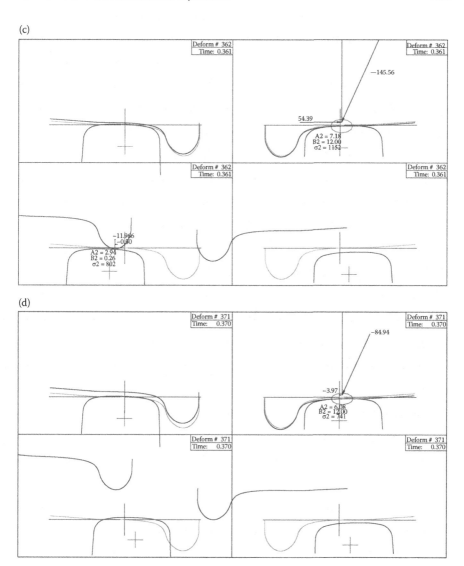

FIGURE 7.20   Continued. Wheel-rail contacts during derailment. (c) $t = 0.36$ s, (d) $t = 0.37$ s.

- Continuous comfort C_{Cx}, C_{Cy} and C_{Cz}. Since the N_{MV}, N_{VA} and N_{VD} methods apply the 95th and 50th percentiles only, there can be substantial losses of useful information concerning isolated events. Therefore, all 5-s rms acceleration values are used in the analysis. These 5-s rms values, called continuous comfort $C_{Cx}(t)$, $C_{Cy}(t)$ and $C_{Cz}(t)$, correspond to the acceleration times series, for x, y and z directions, respectively.
- Comfort on discrete events P_{DE} is used to assess the discomfort due to isolated severe irregularities on straight tracks or on circular curved tracks.

(e)

(f)

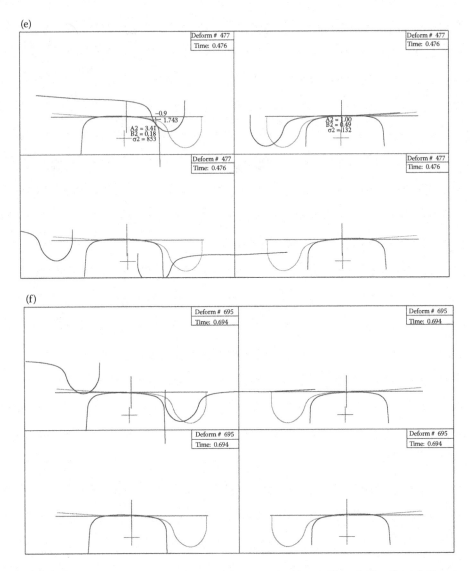

FIGURE 7.20   Continued. Wheel-rail contacts during derailment. (e) $t = 0.47$ s, (f) $t = 0.69$ s.

It is assessed based on the mean lateral acceleration (due to curvature and cant) and the peak-to-peak lateral acceleration. The comfort ranking is given as the percentage of the passengers rating the ride as uncomfortable.

- Comfort on curve transitions P_{CT}, similar to the PDE comfort ranking, is used to assess the discomfort due to the dynamic behaviours experienced on transitions from straight to curved tracks. It is assessed based on the maximum lateral acceleration, the maximum lateral jerk and the maximum roll velocity during transiting the transition. The comfort ranking is given as the percentage of the passengers rating the ride as uncomfortable.

7.7.3 LATERAL INSTABILITY (HUNTING)

Various suitable methods for the lateral instability (hunting) assessment of rail vehicles exist by way of computer analyses of running stability. There are linearised
analyses of the eigenvalues and non-linear simulations which can be used for this
assessment according to various criteria. However, it is important that the lowest
critical speed be found as shown in Figure 7.21.

From Figure 7.21, the wagon speed corresponding to the saddle-node (tangent)
bifurcation point is generally defined as the critical hunting speed, which is usually less
than that corresponding to the subcritical bifurcation. This is usually called the critical
speed in linear analysis at which the dynamic equations have an unstable eigenvalue.

Since the 1980s, non-linear oscillation of wagon hunting has attracted the
attention of several railway researchers. Non-linear stability theories such as the
bifurcation theory were widely applied to analyse the instability due to wagon hunting [41–43]. Several rail vehicle models were developed to determine the critical
bifurcation speed or the location of the turning point in amplitude curves [44–48].
Sensitivity analyses of hunting and chaos in railway lateral dynamics were carried
out [49], and it was concluded that higher wheel-rail friction leads to lower critical speeds, a larger wheelbase leads to higher critical speeds, and, on the 1435 mm
gauge system, the evaluated critical speed significantly increased with the change of
rail cant from 1/20 to 1/40.

An example from [48] is given to show how to determine the critical hunting
speed (or the saddle-node bifurcation) through simulations using a decreasing vehicle speed (e.g. from 200 to 50 km/h) with an initial lateral disturbance for a typical
three-piece bogie of an empty wagon using Gensys software.

A detailed multibody dynamics wagon model is firstly developed using the
Gensys package, as shown in Figure 7.22a and b.

The wagon model in Figure 7.22a includes 11 masses consisting of 1 wagon car
body, 2 bolsters, 4 sideframes and 4 wheelsets, all of which are modelled as rigid
bodies. The connections include a centre bowl and side bearers between the wagon
car body and each bolster, secondary suspensions between each bolster and its sideframes and primary suspensions between each sideframe and its wheelsets. Each
friction wedge is modelled as a massless block and the exact triangular shape is

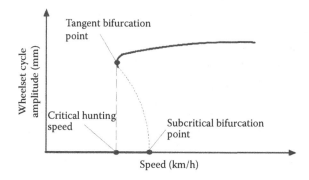

FIGURE 7.21 Bifurcation diagram.

(a)

(b)

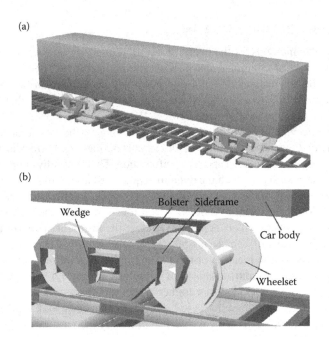

FIGURE 7.22   Three-piece wagon modelling. (a) Whole wagon, (b) bogie.

considered. In the wheel–rail modelling, the Hertzian contact stiffness normal to the wheel–rail contact surface is defined. The wheel–rail contact model allows three different contact surfaces to be in contact simultaneously. The calculations of creep forces are made via a lookup table calculated by Fastsim.

The wheel profiles used for the simulations, and shown in Figure 7.23, are the standard profiles used on the Australian DIRN (defined interstate rail network). The rail profile is as follows:

- *Flat50-new*: A now obsolete 50 kg rail head that featured a very flat head profile and tighter radius on the rail head corners and canted at 1 in 20.

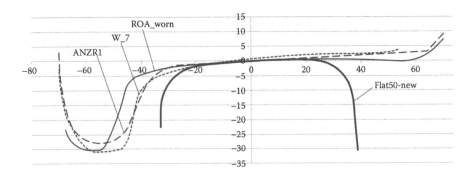

FIGURE 7.23   Flat50-new rail and ANZR1, ROA_worn and W_7 wheel profiles.

The wheel profiles are:

- *ANZR1*: The current new wheel profile for fast freight traffic over the DIRN designed to reduce hunting with lower conicity;
- *ROA (Railways of Australia) Worn*: This is a specific heavily worn hollow wheel profile from the ROA manual used in the hunting test certification until 2009;
- *W_7*: A worn profile with moderate wear at both the tread and the flange contacts, typical of many vehicles.

The equivalent conicities according to UIC Leaflet 519 and the rolling radius differences (rolling radius of the right wheel minus that of the left wheel at nominal gauge) are calculated using Gensys software at the track standard gauge of 1435 mm and are shown in Figure 7.24.

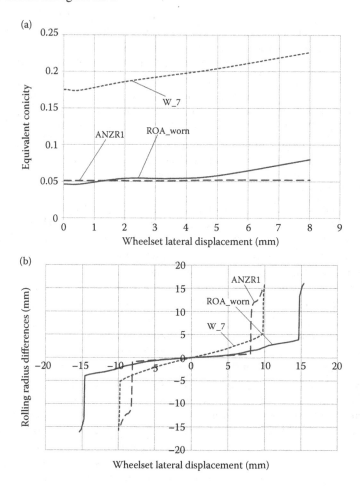

FIGURE 7.24   Wheel–rail contact parameters for Flat50-new rail profile. (a) Equivalent conicities, (b) rolling radius differences.

From Figure 7.24b, it is clearly seen that the increases of rolling radius differences for Flat50-new/W_7 contact as the wheelset lateral displacement increases are much higher than for other contacts, which result in their equivalent conicities being much higher than other contacts, as shown in Figure 7.24a. It is also seen that the change of rolling radius differences for Flat50-new/ANZR1 contact is linear as the wheelset lateral displacement increases. Therefore, the changes of their equivalent conicities are constant. It would be expected that the critical hunting speed of wagons with the Flat50-new/W_7 wheel–rail contact will be much lower than other contacts because they have much higher equivalent conicities.

Simulations in addition to those shown above indicate that, for the contacts of Flat50-new-ANZR1 and Flat50-new-ROA_Worn, the wagon critical hunting speeds are much higher than 150 km/h. As expected, the results for the Flat50-new/W_7 wheel–rail contact give a much lower hunting speed. Therefore, the example for the Flat50-new-W_7 wheel–rail contact at the standard gauge of 1435 mm is investigated in more detail; the time histories of the speed and the lateral displacement of four wheelsets are shown in the upper and lower graphs of Figure 7.25a. As shown in the lower graph of Figure 7.25a, the typical lateral displacement of a wheelset contains a constant limit cycle (close to ±10 mm) from the beginning of the simulation until a certain speed at which the cycle amplitude starts to reduce until disappearing at a slower speed. The latter speed approximates the critical hunting speed and is shown by the solid line in Figure 7.25a, and this speed is about 115.5 km/h. This speed and several higher speeds are used for the constant speed hunting simulations. Therefore, the critical hunting speed can be defined as the speed at which a wagon exhibits a constant limit cycle of wheelset lateral displacement for a sufficiently long time period (e.g. 5 s based on Australian Standard AS7509).

Figure 7.25b shows the simulation results, detailing the time histories of four wheelset lateral displacements at speeds of 115.5, 120.5 and 124.0 km/h. The constant limit cycle disappears after 2 s at the speed of 115.5 km/h, and it disappears after about 2.5 s at the speed of 120.5 km/h and only at the speed of 124.0 km/h does it remain for a period of 5 s. Hence, it can be claimed that the critical hunting speed will be 124.0 km/h, rather than 115.5 km/h. However, in a practical sense, it is recommended that the lower speed be chosen to ensure operational safety.

Figure 7.25c shows the simulation results indicating wheelset hunting phases. It can be seen that the first and second wheelsets in the front bogie always hunt in the same direction and the amplitudes are almost the same, similarly for the third and fourth wheelsets in the rear bogie. The simulations indicate that, at the higher speeds (e.g. >180 km/h), these two bogies hunt out-of-phase, which would cause the wagon car body to have a big yaw rotation, with the maximum lateral acceleration being $0.67g$ at the front and rear bogie centres. At lower speeds (e.g. >150 km/h and <170 km/h), the two bogies hunt in-phase, which would cause the wagon car body to have a big lateral movement, with the maximum lateral acceleration being $0.52g$ at the front and rear bogie centres. At speeds close to the critical hunting speed, these two bogies hunt in a transient stage from in-phase to out-of-phase.

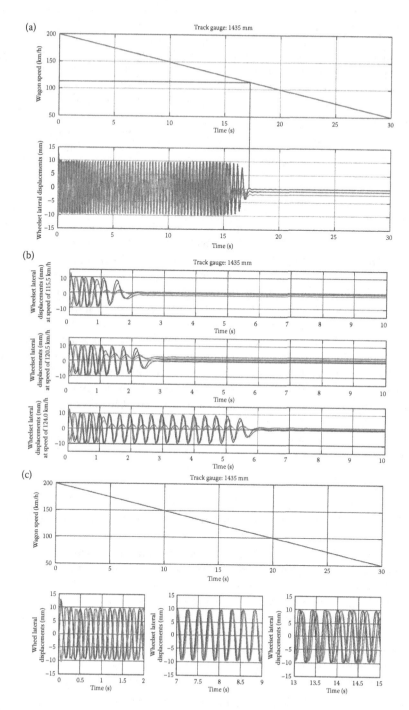

FIGURE 7.25 Results for Flat50-new/W_7 wheel–rail contact. (a) Critical speed determination, (b) critical speed simulations at constant speeds, 115.5; 120.5 and 124.0 km/h, (c) wheelset hunting phases.

7.7.4 CURVING PERFORMANCE

The evaluation and assessment of the curving performance of new or modified rail vehicles can be divided into three categories, these being horizontal and vertical curve negotiation, transition curve negotiation and curving stability. Static or quasi-static analyses can be carried out to investigate the first two categories. However, the curving stability will be evaluated using dynamic analysis.

The purpose of evaluating the first category of horizontal and vertical curve negotiation is to describe how to evaluate whether a vehicle has adequate clearances to enable it to negotiate the tightest horizontal and vertical curves likely to be encountered in the track without uncoupling or doing damage to itself or the track. A longitudinal train simulator (like CRE-LTS introduced in Chapter 5) will be used to simulate the vehicle, which will be coupled to an identical vehicle as well as to vehicles with greater or lesser end throw that are likely to be coupled to it in service, under the track curve conditions as described in Table 7.2.

Based on the simulation results, checks will be undertaken to verify that no unintended contact occurs between components of the vehicle when it is coupled to other vehicles in service when traversing the track geometry scenarios given in Table 7.2. Checks will also be made that adequate clearances exist between the vehicle underframe and its bogies, wheelsets or wheels.

The purpose of the second category is to describe how to evaluate whether a vehicle can safely negotiate the exit transition from curves without the leading wheel flange climbing the rail. A static twist test can be used to assess the wheel unloading performance and underframe behaviour of a rail vehicle on a track geometry that replicates the twist conditions as shown in Figure 7.26. This test can also be

TABLE 7.2
Limiting Curves

	Horizontal Curves			Vertical Curves		
	Simple	Reverse		Convex	Concave	Grade Change
Network/System	Radius (m)	Radius (m)	Length of Intervening Straight (m)	Radius (m)	Radius (m)	Angle (Deg) with No Curve
Australian interstate standard gauge network	90	120	0	300	300	NA
Australasian networks	180			>300	>300	NA
RailCorp Sydney	120	120	20	1300	1300	
Connex. Vic. Electrified 1600 mm gauge	150	150	10	823	305	NA
QR 1067 gauge track	80	80	10	525	300	2.9

Source: From RISSB standard: Dynamic behaviour AS7509.2:2009, Railway rolling stock—Dynamic behaviour—Part 2: Freight rolling stock, Standards Australia, Sydney, Australia, 2009.
Note: NA = not applicable.

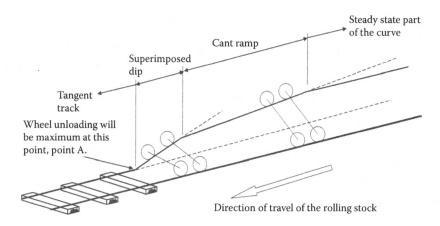

FIGURE 7.26   Twist test.

simulated at a very low speed using one of the software packages introduced in Chapter 5 such as Gensys.

The acceptance criteria include:

- The maximum wheel unloading will not exceed 60%, which is calculated using: $W_{\text{unload}} = 100 \left[1 - \dfrac{\text{minimum wheel load}}{(W_{\text{left\_wheel\_load}} + W_{\text{right\_wheel\_load}})/2} \right]$;

- For rolling stock fitted with a centre plate, the engagement with the bogie bolster centre casting during testing shall not be less than 14 mm.

The third category, curving stability, is concerned about the longitudinal forces in curving performance. The purpose is to describe how to evaluate whether the lateral component of longitudinal train forces during curving will be sufficient to cause wheel lift and subsequent derailment. It is apparent that a dynamic analysis is needed. However, the static analysis is firstly carried out. In the static analysis, the limiting wheel lift ratio R_{WL} for a vehicle should not exceed the limiting value R_{Lim} given in Table 7.3, calculated from the following equation: $R_{WL} = (L_{\text{bogie}}/V_{\text{bogie}}) \leq R_{\text{Lim}} = 0.5 \cdot (G/H)$ in which L_{bogie} is the maximum lateral force resisted by one bogie (kN), V_{bogie} is the sum of the vertical forces per bogie (kN), G is the track gauge (mm) and H is the coupler height above top of rail (mm).

TABLE 7.3

Limiting Wheel Lift Ratios, R_{Lim}

Track Gauge (mm)	Coupler Height (mm)	Limiting Wheel Lift Ratio
1067	785	0.68
1435	876	0.82
1600	876	0.91

It is required that a physical test should be performed if the calculated wheel lift ratio R_{WL} at either bogie is greater than 90% of the limiting ratio R_{Lim}. It is apparent that simulations using professional software packages could substitute such a test. In the dynamic analysis, two kinds of simulation software, longitudinal train simulation and multibody dynamics simulation are required for the evaluation. Hence, the methodology for evaluation and assessment should consist of three interlinked stages:

- Longitudinal train dynamics simulation;
- Identification of the wagon subjected to the maximum lateral coupler forces;
- Wagon dynamics simulation.

An example of such an investigation can be found in Chapter 9 (see Stage 3 in Section 9.2).

7.8   VEHICLE–TRACK INTERACTIONS

The railway transport operation imposes cyclic loading and impact loading on the wheel and rail. Depending on the loading characteristics, contact stress distribution and subsurface stress, plastic deformation and shakedown as well as fatigue crack initiation and propagation, known as rolling contact fatigue (RCF), may occur [51]. This can potentially result in defects such as wheel–rail plastic flow, rail corrugations, squats, wheel tread/rim and railhead shelling and so on. The initiation of fatigue cracks in steel is again a threshold phenomenon. These different mechanisms of deterioration manifest themselves through a number of damage patterns. It has been stated [52] that uncontrolled operationally related wheel–rail damage phenomena mainly included thermal cracks, skidded wheels, skid marks on rails (the so-called wheel-burns) and tread metal build-up on wheels.

7.8.1   INITIATION AND GROWTH OF RCF CRACKS

The magnitude and depth of the maximum wheel–rail contact stress is dependent on the tangential and normal forces and also the curvature of the wheel and rail surfaces. That is why some railway authorities stipulate the wheel–rail impact force limits (such as a P2 force limit [53]) to control the wheel–rail contact stress. High wheel–rail contact stresses will accelerate the deterioration of both wheel and rail through the RCF mechanism.

Depending on the contact conditions, surface cracks can develop into either wear or RCF [54]. Tangential and normal forces, contact pressure, creepage and the presence of fluid are important factors in determining the amount of wear and RCF. Vehicle dynamic modelling undertaken at several rail RCF sites confirmed that the primary causes of RCF damage are large tangential forces and creepages, primarily in the longitudinal (traction) direction along the railhead [55]. It was indicated that the energy dissipation in the contact patch (measured by the product of creep forces and creepages) is the best indicator of the risk of RCF initiation and is also a good indicator of surface crack length.

The contact fatigue stresses experienced by the rail under traction have been shown to be greater than at the wheel or rail at any other time [56]. This is because thermal stresses for the rail in traction add to the maximum shear stresses in the contact patch. Thermal stresses can also contribute to plastic flow and failure in rolling and sliding contact. Contact patch flash temperatures have also been modelled [57], and the significant influence of thermal stresses on the elastic limit and the shakedown limit in a wheel–rail contact are detailed in [58]. The contact patch flash temperature changes the material properties of the steel and, in extreme cases, causes changes to the material crystal structure to form what are known as white etching layers (WELs). The material strength, hardness and elastic modulus all reduce with temperature and are causal to the reduction in friction coefficients [59], affecting fatigue and wear resistance.

RCF stresses increase with wheel rail creep forces [60]. When the creep force reaches a level of approximately 0.3 of the normal force, the maximum shear stress in the wheel–rail contact coincides with the surface as opposed to being at some distance below the surface (usually a distance equal to the contact patch radius). This then makes surface initiated RCF cracks such as head checks and rail squats more prevalent, as shown in Figure 7.27.

Cracks initiated at the wheel surface have been found on certain trains, which has led to wheel damage and more frequent re-profiling, reducing the lifetime of the wheels. For railway wheel applications, RCF can be divided into three different categories:

- Surface-initiated fatigue, sometimes denoted as spalling;
- Subsurface-initiated fatigue, sometimes denoted as shelling;
- Fatigue initiated at deep defects, sometimes denoted as deep shelling or shattered rims.

Studies of RCF defects in rail vehicle wheels, involving wheel spalling, shelling and tread checking [61] showed that the increase in the coefficient of friction (adhesion) resulted in:

- An increase of the stick zone at the contact patch;
- An increase of the maximum value of the principal shear stress;
- A shift of the plastic flow zone from the subsurface to surface for a range of coefficient of friction from 0.35 to 0.4;

FIGURE 7.27   Rolling contact fatigue defects in rail. (a) Head checks, (b) rail squat.

- An increase of the tensile stress with maximum value at the boundary of the contact patch opposite to the direction of movement and at the side boundary as well under lateral creep.

Although a full theoretical understanding of the conditions leading to crack initiation/growth is very complex, a simple analysis of the shakedown diagram (Figure 7.28) and its limits can help in developing RCF reduction criteria as follows:

- Reduction of the wheel–rail maximum contact stress can be achieved either by keeping the area of the contact patch large or by reducing the contact angle at the contact point;
- Reduction of contact stress/traction coefficient product is achieved through the minimisation of longitudinal and lateral contact forces for each wheel.

The potential reduction in RCF can be obtained [62] by installing active bogie steering systems of different bandwidths in an otherwise completely conventional passenger coach. Another method with potential resistance to RCF is the development of bainitic rail steels [63]. Research indicates that bainitic steel rails could have significantly better RCF performance compared to pearlitic steel rails.

7.8.2 Wear of the Rail and Wheel

For this type of wheel damage, the micro-level material removal in combination with plastic deformation may cause both out-of-roundness and uniform profile alterations to the wheel tread fairly constantly around the circumference. Rail damage leaves

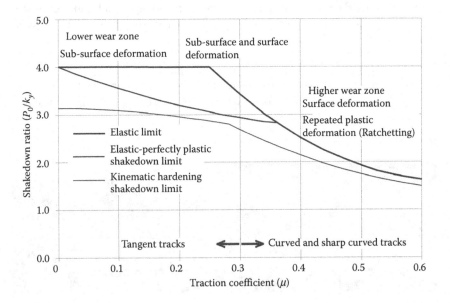

FIGURE 7.28 Shakedown diagram.

longitudinally constant profile-related wear. A non-uniform periodic deformation pattern, known as corrugation, is also common. It was pointed out [51] that the wheel–rail wear is assumed to be proportional to the dissipated energy or the friction power. The important parameters affecting wear rates mainly included friction, normal load and sliding velocity.

To assess the wheel–rail damage, two indices should be introduced. One is the wear index based on Burstow's modelling [64] using the creep energy $T\gamma$ (creep force multiplied by creepage) and widely used in VAMPIRE software [65]. For mild wear, $T\gamma < 160$, and for severe wear, $T\gamma \geq 160$. The other is the surface fatigue index defined by Ekberg et al. [66,67] based on the shakedown diagram shown in Figure 7.28.

Tangential and normal forces, contact pressure, creepage and the presence of fluid are important factors in determining the amount of wear and RCF. There are two types of wear dominant in wheel–rail contact, adhesive and delamination, and there are two basic types of wear models for wheel–rail interaction, energy transfer models and sliding models.

To investigate the load at which continuous plastic flow begins, that is, to determine the shakedown limit [68], it was found with twin-disc tests that macroscopic plastic deformation starts at contact pressures above 1000 MPa in pure rolling, and at 750 MPa when combined with a tangential load corresponding to a traction coefficient of 0.3. The shakedown limit was estimated as 1600 MPa in pure rolling and 1200–1300 MPa with a traction coefficient of 0.25 and nominal yield strength. The explanation given in [69] for the generation of plastic flow with an operating contact pressure below the shakedown limit is that this is due to surface roughness, which leads to increases in peak pressures by as much as a factor of 8. ABAQUS FE rail modelling [70] was used to simulate both local elastic and plastic deformations as the wheel rolls along the rail, with a stress–strain curve having 2% plastic strain at a stress of 730 MPa, 5% at 920 MPa and 20% at 1000 MPa.

Although corrugation occurs from different mechanisms in different places, one effective treatment [56] is to control wheel–rail friction. Friction is potentially significant for all types of corrugation in which wear is damage mechanism. If the rail-head friction coefficient can be controlled to a value of about 0.35, this is sufficient for vehicle traction and braking while limiting damage from plastic flow and wear corrugation.

The thermal-traction wheel–rail defects mainly include wheel flats, wheel (or engine) burns and rail squats. These defects are associated with a combination of thermal, traction (and braking), creep and slippage effects that can develop at the wheel–rail interface in any track section, and under all types of operating conditions (ranging from passenger to heavy haul). Wheel flat defects are caused by extreme train braking, and wheel (or engine) burn defects are caused by the continuous slipping of the locomotive or traction wheels on the rails that occurs when the longitudinal creepage between wheels and rails reaches saturation. Squat defects have been extensively investigated recently. Squat defects are initiated primarily by surface damage that can occur due to excessive wheel–rail creep forces, leading to the exhaustion of the rail material's ductility, or the transformation of the material to a 'white etching' brittle layer, or the extension of existing gauge corner checking defects [71–73].

7.8.3 SHAKEDOWN DIAGRAM

The shakedown diagram shown in Figure 7.28 is mainly used to examine the influence of material strength on damage at the wheel–rail interface [74]. The diagram is plotted by taking into consideration traction or creepage levels at the wheel–rail interface (X-axis), and the maximum contact stress divided by the yield strength of the rail material in shear (Y-axis), which is designated as the shakedown ratio. In this way, the shear stress in the contact patch is measured. It is believed that, when the effective shear stress exceeds the yield strength, then the material will plastically deform and start ratchetting. In Figure 7.28, at certain axle loads, the shakedown ratio for standard carbon rail is well above the kinematic hardening shakedown limit. Under this circumstance, repeated plastic deformation (ratchetting) will occur, resulting in surface and sub-surface cracks developing into either wear or RCF defects. By using head hardened rail [75] at similar axle loads, the shakedown ratio reduces to the elastic shakedown region where the performance of rail will stabilise and behave more or less elastically.

It can be seen from Figure 7.28 that, above 0.3 traction/creepage coefficient, both the kinematic hardening shakedown limit and the elastic limit decrease significantly as the traction/creepage coefficient increases because high traction/creepage effects introduce high shear stress near the rail surface, which will result in the development of either wear or RCF defects. There are two situations which can increase the traction/creepage coefficient—one is the introduction of higher adhesion locomotives in the railway networks, and the other is train operation on sharp curved tracks.

In [66,67], the surface fatigue index FI_{surf} was defined based on the shakedown map, as shown in Figure 7.28. $FI_{surf} = T/N - k_y/P_0$ in which $\mu = T/N$ is the traction coefficient, k_y is the yield stress in pure shear and P_0 is the maximum contact stress. For $\mu > 0.3$, the first yielding occurs at the surface for progressively smaller maximum contact stresses as the traction coefficient increases.

7.9   VEHICLE ACCEPTANCE SIMULATIONS

The acceptable dynamic behaviours of rail vehicles are governed by several standards around the world. Some standards allow the use of multibody simulation tools (such as VAMPIRE, NUCARS, GENSYS and SIMPACK) in place of physical testing, but generally not for all vehicle tests within each standard. Virtual multibody vehicle models can allow simple analyses, such as for slightly modified vehicles, to be completed with less time, cost and effort in comparison to physical testing. Unfortunately, the detailed vehicle model acceptance procedures required to achieve this for new rail vehicle designs do not presently exist. However, the methodology behind the proposed locomotive model acceptance procedure (LMAP) [76] and wagon model acceptance procedure (WMAP) [77] currently intended for Australian freight vehicles is described as follows, although it can be modified to suit other countries and vehicle types:

- Create a vehicle model acceptance procedure;
- Incorporate the best practices from international and local standards applicable to vehicle dynamic behaviour, including the traction and braking tests in cases where these are required;

- Traction and dynamic braking power controls have been added in the multibody code as subroutines;
- Simulate the whole model using the new vehicle model acceptance procedure.

Locomotive and wagon model acceptance procedures have been proposed in [76,77] to validate the multibody models of vehicles intended for use on Australian railways. Although largely based on the RISSB/Australian Standards, some content from other Australian and worldwide standards, along with various MBS software manuals, was selected to augment tests within the procedure. The case studies selected show how the proposed vehicle model acceptance procedures can be implemented when testing locomotive models for validation purposes. The advantage of this approach is that it improves the likelihood of identifying errors within vehicle models that could otherwise go unnoticed. Once all tests in the proposed procedure have been performed on the vehicle model, data from these simulations can be compared to experimental data from an equivalent real-world vehicle. This offers further scope to improve the model to the point where it can accurately and comprehensively replicate the dynamic behaviour of its physical counterpart whilst still satisfying the requirements of relevant standards.

REFERENCES

1. J.J. Kalker, A fast algorithm for the simplified theory of rolling contact, *Vehicle System Dynamics*, 11(1), 1982, 1–13.
2. V.K. Garg, R.V. Dukkipati, *Dynamics of Railway Vehicle Systems*, Academic Press, New York, NY, 1984.
3. R.V. Dukkipati, J.R. Amyot, *Computer-Aided Simulation in Railway Dynamics*, Marcel Dekker, New York, NY, 1988.
4. K.L. Johnson, *Contact Mechanics*, Cambridge University Press, Cambridge, UK, 1985.
5. A. Cai, G.P. Raymond, Modelling the dynamic response of railway track to wheel/rail impact loading, *Structural Engineering and Mechanics*, 2(1), 1994, 95–112.
6. A. Cai, G.P. Raymond, Theoretical model for dynamic wheel/rail and track interaction, *Proceedings of the 10th International Wheelset Congress*, Sydney, Australia, 27 September–1 October 1992, pp. 127–131.
7. A. Igeland, Railhead corrugation growth explained by dynamic interaction between track and bogie wheelsets, *Rail and Rapid Transit*, 210(1), 1996, 11–20.
8. G. Lu, K.F. Gill, Track–transmission system dynamic analysis, *Rail and Rapid Transit*, 207(2), 1993, 99–113.
9. C. Andersson, J. Oscarsson, Dynamic train/track interaction including state-dependent track properties and flexible vehicle components, *Vehicle System Dynamics*, 33(Supp), 2000, 47–58.
10. T. Dahlberg, Vertical dynamic train/track interaction—Verifying a theoretical model by full-scale experiments, *Vehicle System Dynamics*, 24(Suppl), 1995, 45–57.
11. Y.Q. Sun, M. Dhanasekar, A dynamic model for the vertical interaction of the rail track and wagon system, *International Journal of Solids and Structures*, 39(5), 2002, 1337–1359.
12. Y.Q. Sun, M. Dhanasekar, D. Roach, A 3D model for the lateral and vertical dynamics of wagon–track system, *Rail and Rapid Transit*, 217(1), 2003, 31–45.
13. Y.Q. Sun, M. Dhanasekar, Importance of track modelling to the determination of the critical speed of wagons, *Vehicle System Dynamics*, 41(Suppl), 2004, 232–241.

14. Y.Q. Sun, C. Cole, Comprehensive wagon-track modelling for simulation of three-piece bogie suspension dynamics, *Journal of Mechanical Engineering Science*, 221(8), 2007, 905–917.
15. Y.Q. Sun, S. Simson, Wagon-track modelling and parametric study on rail corrugation initiation due to wheel stick–slip process on curved track, *Wear*, 265(9–10), 2008, 1193–1201.
16. Y.Q. Sun, C. Cole, Vertical dynamic behaviour of three-piece bogie suspensions with two types of friction wedge, *Journal of Multibody System Dynamics*, 19(4), 2008, 365–382.
17. D.R. Ahlbeck, H.C. Meacham, R.H. Prause, The development of analytical models for railroad track dynamics, in *Railroad Track Mechanics & Technology: Proceedings of a Symposium held at Princeton University in April 1975*, A.D. Kerr (Ed.), Pergamon Press, New York, NY, 1978, pp. 239–263. ISBN: 9780080219233.
18. W. Zhai, X. Sun, A detailed model for investigating interaction between railway vehicle and track, *Vehicle System Dynamics*, 23(Suppl), 1994, 603–615.
19. S.L. Grassie, R.W. Gregory, D. Harrison, K.L. Johnson, The dynamic response of railway track to high frequency vertical excitation, *Journal of Mechanical Engineering Science*, 24(2), 1982, 77–90.
20. G. Diana, F. Cheli, S. Bruni, A. Collina, Interaction between railroad superstructure and railway vehicles, *Vehicle System Dynamics*, 23(Suppl), 1994, 75–86.
21. L.M. Martin, J.G. Gimenez, Railway vehicle modelling by the constraint equation method, *Vehicle System Dynamics*, 13(5), 1984, 281–297.
22. M. Jochim, W. Kortum, G. Kocher, Analysis of railway vehicle system dynamics with the multibody program MEDYNA, in *Computer Applications in Railway Operations*, T.K.S. Murthy, B. Mellitt, S. Lehmann, G. Astengo (Eds.), Computational Mechanics Publications, Southampton, UK, 1990, pp. 189–208. ISBN: 9780945824411.
23. C.F. Gerald, P.O. Wheatly, *Applied Numerical Analysis* (5th ed.), Addison-Wesley, Reading, MA, 1994.
24. R.L. Burden, J.D. Faires, *Numerical Analysis* (6th ed.), Brooks/Cole, Pacific Grove, CA, 1997.
25. W. Zhai, Locomotive–track system coupling dynamics and its application to the study of locomotive performance, *Journal of China Railway Science (in Chinese)*, 17(2), 1996, 58–73.
26. M. Miyamoto, Mechanism of derailment phenomena of railway vehicles, *Quarterly Report of Railway Technical Research Institute, Japan*, 37(3), 1996, 147–155.
27. H. Ishida, M. Matsuo, Safety criteria for evaluation of railway vehicle derailment, *Quarterly Report of Railway Technical Research Institute, Japan*, 40(1), 1999, 18–25.
28. A. Matsuura, Dynamic interaction of vehicle and track, *Quarterly Report of Railway Technical Research Institute, Japan*, 33(1), 1991, 31–38.
29. Y.Q. Sun, C. Cole, M. Dhanasekar, D.P. Thambiriratnam, Modelling and analysis of the crush zone of a typical Australian passenger train, *Vehicle System Dynamics*, 50(7), 2012, 1137–1155.
30. Y.Q. Sun, N. Zong, M. Dhanasekar, C. Cole, Dynamic analysis of vehicle-track interface under train collision using multi-body dynamics, *Proceedings of the 23rd International Symposium on Dynamics of Vehicles on Roads and on Tracks*, 19–23 August 2013, Qingdao, China.
31. J. Förstberg, Ride comfort and motion sickness in tilting trains: Human responses to motion environments in train experiment and simulator experiments, PhD thesis, Department of Vehicle Engineering, Royal Institute of Technology, Stockholm, Sweden, 2000.
32. J. Förstberg, *Human Responses and Motion Environments in a Train Experiment (SJ X2000)*, Swedish National Road and Transport Research Institute, Linköping, Sweden, 2000.

33. BS EN 12299:2009, *Railway Application—Ride Comfort for Passengers—Measurement and Evaluation*, British Standards Institution, London, 2009.

34. B. Kufver, R. Persson, J. Wingren, Certain aspects of the CEN standard for the evaluation of ride comfort for rail passengers, *in WIT Transactions on the Built Environment*, 114, 2010, 605–614.

35. C.D. Ketchum, N. Wilson, *Transit Cooperative Research Program Web-Only Document 52: Performance-Based Track Geometry Phase 1*, Transportation Technology Centre Inc., Pueblo, CO, 2012.

36. ISO 2631–2:2003, *Mechanical Vibration and Shock—Evaluation of Human Exposure to Whole Body Vibration—Part 2: Vibration in Buildings (1 Hz to 80 Hz)*, International Organization for Standardization, Geneva, Switzerland, 2003.

37. RDS 7513.3, *Railway Rolling Stock—Interior Environment—Part 3: Passenger Rolling Stock, Draft 2.1*, Rail Industry Safety and Standards Board, Canberra, Australia, 2010.

38. Y. Sugahara, A. Kazato, R. Koganei, M. Sampei, S. Nakaura, Suppression of vertical bending and rigid-body-mode vibration in railway vehicle car body by primary and secondary suspension control: Results of simulations and running tests using Shinkansen vehicle, *Rail and Rapid Transit*, 223(6), 2009, 517–531.

39. T. Tamioka, T. Takigami, Reduction of bending vibration in railway vehicle carbodies using carbody–bogie dynamic interaction, *Vehicle System Dynamics*, 48(Suppl), 2010, 467–486.

40. R.C. Sharma, Parametric analysis of rail vehicle parameters influencing ride behaviour. *International Journal of Engineering, Science and Technology*, 3(8), 2011, 54–65.

41. X. He, R.R. Huilgol, Application of Hopf bifurcation at infinity to hunting vibrations of rail vehicle trucks, *Vehicle System Dynamics*, 20(Suppl), 1992, 240–253.

42. H. True, Railway vehicle chaos and asymmetric hunting, *Vehicle System Dynamics*, 20(Suppl), 1991, 625–637.

43. M. Rose, H. True, Nonlinear dynamics of railway vehicles, *Nonlinear Science Today*, 3, 1993, 1–4.

44. G. Xu, A. Steindl, H. Troger, Nonlinear stability analysis of a bogie of a low-platform wagon, *Vehicle System Dynamics*, 20(Suppl), 1991, 653–665.

45. H. Chengrong, Z. Feisheng, The numerical bifurcation method of nonlinear lateral stability analysis of a locomotive, *Vehicle System Dynamics*, 23(Suppl), 1994, 234–245.

46. J. Zeng, Numerical computations of the hunting bifurcation and limit cycles for railway vehicle system (in Chinese), *Journal of China Railway Society*, 18(3), 1996, 13–19.

47. S. Simson, Y.Q. Sun, C. Cole, Bogie warp and friction interaction with effective conicity on the hunting of three-piece bogies, *Proceedings of the 22nd International Symposium on Dynamics of Vehicles on Roads and on Tracks (CD)*, Manchester, UK, 14–19 August 2011.

48. Y.Q. Sun, M. Spiryagin, C. Cole, S. Simson, Effect of wheel–rail contacts and track gauge variation on hunting behaviours of Australian three-piece bogie wagon, *Proceedings of the 23rd International Symposium on Dynamics of Vehicles on Roads and on Tracks*, Qingdao, China, 19–23 August 2013.

49. H. True, J.C. Jensen, Parameter study of hunting and chaos in railway vehicle dynamics, *Vehicle System Dynamics*, 23(Suppl), 1994, 508–521.

50. AS7509.2:2009, Railway rolling stock—Dynamic behaviour—Part 2: Freight rolling stock, Standards Australia, Sydney, Australia, 2009.

51. R.D. Fröhling, Wheel/rail interface management in heavy haul railway operations—Applying science and technology, *Vehicle System Dynamics*, 45(7–8), 2007, 649–677.

52. D.R. Ahlbeck, L.E. Daniels, Investigation of rail corrugations on the Baltimore Metro, *Wear*, 144(1–2), 1991, 197–210.

53. AS7508.2:2008, *Railway Rolling Stock—Track Forces & Stresses—Part 2: Freight Rolling Stock*, Standards Australia, Sydney, Australia, 2008.

54. J. Tunna, J. Sinclair, J. Perez, A review of wheel wear and rolling contact fatigue, *Rail and Rapid Transit*, 221(2), 2007, 271–289.

55. M.J.M.M. Steenbergen, Modelling of wheels and rail discontinuities in dynamic wheel–rail contact analysis, *Vehicle System Dynamics*, 44(10), 2006, 763–787.

56. S. L. Grassie, Rail corrugation: Characteristics, causes and treatments, *Rail and Rapid Transit*, 223(6), 2009, 581–596.

57. M. Ertz, K.A. Knothe, Comparison of analytical and numerical methods for the calculation of temperatures in wheel/rail contact, *Wear*, 253(3–4), 2002, 498–508.

58. J.R. Evans, M.C. Burstow, Vehicle/track interaction and rolling contact fatigue in rails in the UK, *Vehicle System Dynamics*, 44(Suppl), 2006, 708–717.

59. C. Tomberger, P. Dietmaier, W. Sextro, K. Six, Friction in wheel–rail contact: A model comprising interfacial fluids, surface roughness and temperature, *Wear*, 271(1–2), 2011, 2–12.

60. O. Polach, Creep forces in simulations of traction vehicles running on adhesion limit, *Wear*, 258(7–8), 2005, 992–1000.

61. P.A. Meehan, W.J.T. Daniel, T. Campey, Prediction of the growth of wear-type rail corrugation, *Wear*, 258(7–8), 2005, 1001–1013.

62. H.H. Jenkins, J.E. Stephenson, G.A. Clayton, G.W. Morland, D. Lyon, The effect of track and vehicle parameters of wheel/rail vertical dynamic forces, *Railway Engineering Journal*, 3(1), 1974, 2–16.

63. K. Sawley, J. Kristan, Development of bainitic rail steels with potential resistance to rolling contact fatigue, *Fatigue & Fracture of Engineering Materials & Structures*, 26(10), 2003, 1019–1029.

64. M. Burstow, Whole life rail model application and development for RSSB: Development of an RCF damage parameter, Report Number AEATR-ES-2003-832 Issue 1, Rail Safety & Standards Board, London, UK, 2003.

65. L. Rawlings (Ed.), *VAMPIRE (Version 4.32) User Manual*, AEA Technology, Derby, UK, 2004.

66. A. Ekberg, E. Kabo, H. Andersson, An engineering model for prediction of rolling contact fatigue of railway wheels, *Fatigue & Fracture of Engineering Materials & Structures*, 25(10), 2002, 899–909.

67. A. Ekberg, E. Kabo, Fatigue of railway wheels and rails under rolling contact and thermal loading—An overview, *Wear*, 258(7–8), 2005, 1288–1300.

68. H. Krause, G. Poll, Plastic deformations of wheel–rail surfaces, *Wear*, 113(1), 1986, 123–130.

69. A. Kapoor, F.J. Franklin, S.K. Wong, M. Ishida, Surface roughness and plastic flow in rail wheel contact, *Wear*, 253(1–2), 2002, 257–264.

70. M. Busquet, H. Chollet, L. Baillet, C. Dagorn, J.B. Ayasse, Y. Berthier, From railway dynamics to wheel/rail contact mechanics, an approach for the modelling of the wheel/rail contact: Elasto-plastic response of the railhead, *Rail and Rapid Transit*, (3), 2006, 189–200.

71. S. Marich, Practical/realistic implementation of wheel/rail contact technologies—The Australian experience, *Proceedings of the 7th International Conference on Contact Mechanics and Wear of Rail/Wheel Systems*, 24–27 September 2006, Brisbane, Australia, pp. 3–22.

72. S. Marich, S. Mackie, The development of squat defects under high axle load operations, *Proceedings of the Conference on Railway Engineering*, Wollongong, Australia, 10–13 November 2002, pp. 17–26.

73. M. Kerr, A. Wilson, S. Marich, The epidemiology of squats and related rail defects, *Proceedings of the Conference on Railway Engineering*, Perth, Australia, 7–10 September 2008, pp. 83–96.

74. M. Ertz, K. Knothe, Thermal stresses and shakedown in wheel/rail contact, *Archive of Applied Mechanics*, 72(10), 2003, 715–729.

75. D.R. Welsby, P.J. Mutton, Influence of rail material grade on wear and rolling contact fatigue behaviour, *Track & Signal*, 17(1), 2013, 70–71.

76. M. Spiryagin, A. George, Y.Q. Sun, C. Cole, T. McSweeney, S. Simson, Investigation of locomotive multibody modelling issues and results assessment based on the locomotive model acceptance procedure, *Rail and Rapid Transit*, 227(5), 2013, 453–468.

77. M. Spiryagin, A. George, S. Ahmad, K. Rathakrishnan, Y. Sun, C. Cole, Wagon model acceptance procedure using Australian Standards, *Proceedings of the Conference on Railway Engineering*, Brisbane, Australia, 10–12 September 2012, pp. 343–350.

8 Co-Simulation and Its Application

8.1 INTRODUCTION TO CO-SIMULATION PROCESS

The development of new rail vehicles or the investigation of the vehicle dynamic behaviour of existing vehicles requires the application of an advanced modelling methodology because a rail vehicle is a complex system which includes not only mechanical, but also electrical, hydraulic and other subsystems. In a real vehicle, all systems should 'communicate' between themselves. However, in the simulation world it is quite difficult to find a software product which can work in multi-disciplinary areas. Therefore, it is necessary to have different software products for each discipline and to allow them to talk to each other. Some ideas on how this can be organised have been published by Körtum and Vaculín [1]. The simplified scheme for the multi-disciplinary simulation is shown in Figure 8.1. Moreover, not only communication needs to be achieved, but different software packages should also be synchronised in time in order to provide accurate results. There are few cases where time-independent parameters are in use. All these issues are included in the advanced simulation methodology which can be covered by the co-simulation process. Here is a reasonable question: What does co-simulation mean? Let us see what definitions are commonly used for co-simulation.

Co-simulation has been defined in several papers and documents. Arnold et al. [2, pp. 27–28] define a co-simulation approach as one in which 'the subsystems are handled or integrated by different integration methods or solvers and each of these methods or solvers can be or is tailored to the corresponding subsystems'. In addition, it should be equipped with a data exchange algorithm which has discrete synchronisation points with a sampling time of 0.001 s for vehicle system dynamics studies [3]. Another definition can be found in [4, p. 17], which states that 'co-simulation is used to solve a coupled system by simulating each part with its own coupleable simulation tool'. A further comprehensive definition is presented in [5, p. 76] as 'co-simulation exploits the modular structure of coupled problems in all stages of the simulation process beginning with the separate model setup and pre-processing for the individual subsystems in different simulation tools'.

From an engineering point of view, it is better to define a co-simulation process as a simulation process of the whole system, where two or more subsystems are connected between each other in one simulation environment by specialised communication interface(s) with a pre-defined time step for data exchange.

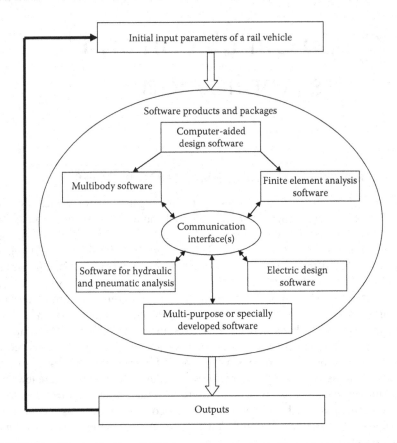

FIGURE 8.1   The application of different software products for computer simulation in the rail vehicle design process.

In common practice, the data exchange process can be achieved through three types of communication techniques:

- Integrated memory-shared communication between software products;
- Network data exchange;
- Exporting code from one package to another.

All of these techniques have their own advantages and disadvantages; the decision about which is best to use is usually based on the initial requirements and existing hardware or software limitations.

8.2   CO-SIMULATION BETWEEN MULTIBODY SOFTWARE PACKAGES AND MATLAB®/SIMULINK®

Simulink® software included in the MATLAB® package is a very powerful tool which allows the development of a model of a full mechatronic system and the simulation of its dynamic behaviour. Simulink also includes a lot of libraries for the design of mechanical, electrical, hydraulic subsystems and so forth that makes the design process reasonably uncomplicated and user friendly. Some rail vehicle mechanical models have been implemented in Simulink [6,7] and simulation results confirm the possibility of the application of Simulink as a stand-alone development tool. However, there are some restrictions and limitations present in comparison with specialised multibody software packages, including:

- The necessity to develop a model from the zero point;
- The absence of standard input parameters and libraries required for railway vehicle design;
- The development of post-processing analysis is required for the models.

Therefore, it is better to use Simulink in co-simulation mode with specialised multibody codes for rail vehicle dynamics studies because this product readily allows the creation of additional subsystems, parts and elements which either cannot be developed inside multibody codes, or their development inside such packages is highly complicated for end-users. In the following subsections, an overview of how the co-simulations are achieved between Simulink and commonly used rail vehicle software packages is presented and some advantages and disadvantages are discussed.

8.2.1 SIMPACK AND SIMULINK®

Regarding the documentation provided by Simpack [8–10], this product supports three types of communication techniques:

- SIMAT interface (matrix data exchange or network data exchange);
- CodeExport (exporting code from Simpack to Simulink);
- MatSIM (exporting code from Simulink to Simpack).

The SIMAT interface allows communication with different software packages during a simulation process. One of these packages is Simulink. The data exchange can be realised through a model linearisation process and presenting a multibody model as the linear system matrix, which can be implemented and saved in the MATLAB m-file. This file contains data required for matrices which can be used for the State-Space function block in Simulink. This approach is not recommended for a real vehicle dynamics analysis, so it is better to use SIMAT's network data exchange. The approach is based on data exchange through the TCP/IP (Transmission Control Protocol/Internet Protocol) protocol as shown in Figure 8.2. It is necessary to define input and output parameters of the multibody model inside the Simpack environment. The network co-simulation interface is realised as the S-Function block in Simulink. The network addresses and ports required for the communication should be defined in both systems. This approach is a very good tool for vehicle system dynamics and is based on the synchronisation method with discrete time points. Its application does show two limitations: that only ODE (ordinary differential equations) solvers can be used in the co-simulation process for the present Simulink versions and only one Simpack co-simulation interface can be used in a Simulink model [8]. Examples of

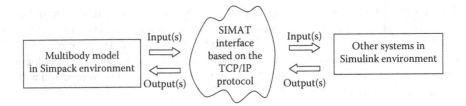

FIGURE 8.2   The co-simulation process between Simpack and Simulink with the network data exchange interface (SIMAT).

applications of such a communication interface for rail vehicle dynamics analysis can be found in [11–13].

CodeExport allows the export of models from Simpack and generation of a source code in Fortran, which can then be converted into C language. The obtained code can be used in Simulink as an S-Function or can find further use in the implementation of software-in-the-loop, hardware-in-the-loop and real-time simulations. This approach has a lot of limitations and restrictions. For example, a great number of force elements are not supported by CodeExport. In the case of simulations for rail vehicles, the existing wheel–rail contact models are not supported when using this approach. However, it is still possible to use this process as shown in Figure 8.3. An example of such an application technique can be found in [14].

MatSIM [10] allows the creation of a Dynamic Link Library (DLL) from a Simulink model by means of the application of Real-Time Workshop (in the latest version of MATLAB–Simulink Coder [15]). The special configuration files for such a process are supplied with the Simpack package. The obtained dll file is then used by Simpack as a plug-in. The process of this technique is presented in Figure 8.4. After that, the new Simpack Control Element can be used in the Simpack model.

To summarise, it is possible to say that Simpack is a very powerful tool for multibody simulations and has a great number of special features for the simulation of complex dynamic systems for rail vehicles. However, the further improvement and development of enhancements will be very useful for rail vehicle dynamicists.

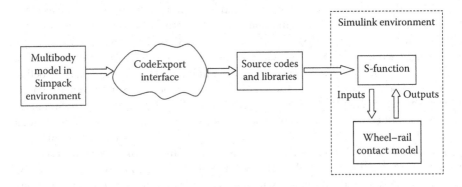

FIGURE 8.3   The implementation of CodeExport for rail vehicle dynamics simulation.

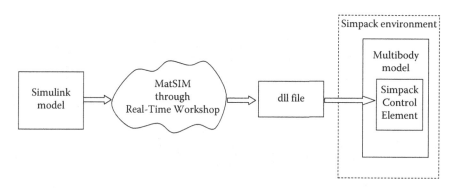

FIGURE 8.4   The implementation of MatSIM for the co-simulation process between Simulink and Simpack.

8.2.2   VI-Rail (ADAMS/Rail) and Simulink

One of the leading and first products in the field of multibody dynamics is MSC. ADAMS (Automatic Dynamic Analysis of Mechanical Systems). In 1995, the special plug-in called Adams/Rail was developed for the study of rail vehicle dynamics. In the middle of 2000, Adams/Rail was replaced by the new plug-in called VI-Rail produced by the company VI-grade. VI-Rail is based on the same algorithms, structures and co-simulation principles as its predecessor. The co-simulation between VI-Rail and Simulink is performed by means of the additional plug-in called Adams/ Controls [16,17]. This plug-in supports two co-simulation techniques mentioned in Section 8.1, these being:

- Exporting code from one package to another;
- Network data exchange with TCP/IP.

The export of code can be done in two ways. The first one is from MSC.ADAMS to Simulink by means of exporting a mechanical model into the MATLAB function. The second way is to import a control system model from Simulink to MSC.ADAMS. In this case, the model should be exported by means of Real-Time Workshop (in the latest version of MATLAB–Simulink Coder [15]) into the file written in C language and then combined with a mechanical model in Adams for further simulation processing.

For the network data exchange, Adams runs as a server and MATLAB/Simulink as a client (see Figure 8.5). In Simulink, the adjustment of simulation parameters can be done through the Function Block Parameters interface developed by MSC.ADAMS.

Before starting simulations, all input and output parameters should be defined in models, depending on what simulation technique is going to be used. In addition, both continuous and discrete modes can be used for the integration of ADAMS and control models. However, for most simulations, it is recommended to use the discrete mode. The continuous mode can be used in some cases when a small time step is required [16].

Some examples of co-simulation application between Adams/Rail and Simulink by means of Adams/Controls can be found in [18–21].

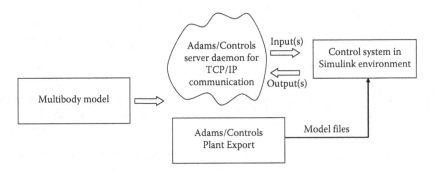

FIGURE 8.5 The co-simulation process between MSC.ADAMS and Simulink using TCP/IP.

8.2.3 VAMPIRE AND SIMULINK

The communication technique between Vampire and Simulink [22] is performed with a specially developed interface, which is implemented as an S function in Simulink and written as an m-file. This function calls a required command from the DLL developed by Vampire (see Figure 8.6). This interface allows users to use a co-simulation process for the development of their own suspension systems, but has a limitation in that it allows the Simulink model to use only one Vampire communication function inside the model. When the Simulink model runs, it calls the Vampire function and Vampire Control analysis is fully controlled by Simulink. The process flow chart is shown in Figure 8.7.

The Vampire package provides several examples of Simulink models for rail vehicle suspension design with a co-simulation process such as a spring element design, a displacement control for a tilting vehicle and so forth [22].

8.2.4 UNIVERSAL MECHANISM AND SIMULINK

Universal Mechanism supports two co-simulation techniques [23,24]:

- Exporting code from Universal Mechanism to Simulink (UM/CoSimulation);
- Exporting code from Simulink to Universal Mechanism (UM/Control).

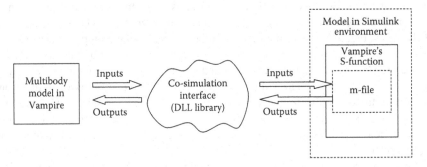

FIGURE 8.6 The co-simulation process between Vampire and Simulink.

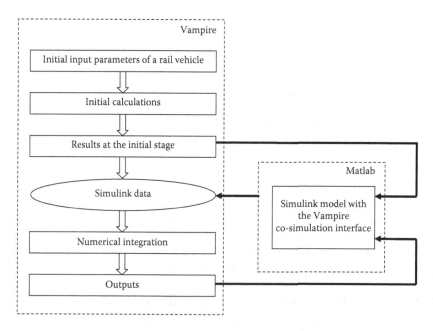

FIGURE 8.7   The flow chart for the co-simulation process between Vampire and Simulink.

The first one generates an m-file and a data file with information about model inputs/outputs. In Simulink, the S-function is used for the co-simulation process and this function works with the m-file created during the model export. The architecture of such a technique is very similar to Vampire, but the difference is that it allows the use of several multibody models exported from Universal Mechanism inside the Simulink model. This means that each of the models has its own m-file. During a simulation process, Simulink creates an m-file, which contains function calls for three stages: initialisation, calculation of output values and termination.

The second technique, which involves exporting code from Simulink and is called MATLAB Import in the Universal Mechanism documentation, is very similar to the MatSIM interface in Simpack. The Real-Time Workshop is used in this case. For each MATLAB software version, specific setting files are required. As a result, the Real-Time Workshop generates a dll file. Subsequently, the library of dll files obtained for all of the versions should be connected to the Universal Mechanism software as an external library. Some limitations are present for such a technique: only parameters which are parts of force element parameters can be used as inputs for the multibody model in Universal Mechanism and outputs of the Simulink model, respectively.

Examples of applications of co-simulation processes between Universal Mechanism and Simulink can be found in [25–27].

8.3   DESIGN OF THE CO-SIMULATION INTERFACE

Summarising the analysis results of co-simulation techniques, it is possible to state that the best solution for co-simulation is the one which is based on network data

FIGURE 8.8   The basic architecture for the co-simulation process with TCP/IP.

exchange, because it avoids some limitations provided by third-party software components (e.g. Real-Time Workshop limitations for a model generating process in Simulink—not all functions and toolboxes are supported). For the network communication, it is better to use a TCP/IP network connection because it provides a controlled data exchange process.

It is common practice for the client–server architecture to be used for the co-simulation process between a multibody software product and Simulink. The basic principle of such architecture is shown in Figure 8.8.

In this chapter, the same approach is used for the development of the co-simulation interface between Gensys and Simulink software products.

8.3.1 Description of the Co-Simulation Interface in GENSYS

Starting from version 10.10, Gensys provides a co-simulation interface and works as a server. In order to activate a co-simulation mode, the following string should be inserted in the 'tsimf' model file:

cosim_server 51717

where 51717 is a port number. This value can be changed by a user, and it is recommended to use values from 49152 to 65535, the range of port numbers not controlled by the Internet Assigned Numbers Authority.

Gensys's server, which can be called 'Server', supports the following commands [28]:

- 'ask_iadr *<string_var_name>*'—this command is needed to obtain the address of a variable. When such a command is sent, the Server sends a response with the address of the variable.
- 'get_iadr *<var_address>*'—this command is needed to obtain the value of a variable. The is defined with the 'ask_iadr' command. As a result of the 'get_iadr' command, the Server sends a response with the value of the variable.
- 'put_iadr *<var_address_value>*'—this command is needed to overwrite the value of a variable. The *<var_address_value>* is also defined with the 'ask_iadr' command. When the 'put_iadr' command is successful, the Server sends a response with the zero value.
- 'run_tout'—this command is required to start a simulation process.
- 'run_stop'—this command is required to stop a simulation process and close a network connection.

If some errors occur during the communication process, the Server can send the following error codes as a response: 7FFFFFFFFFFFFFF6, 7FFFFFFFFFFFFFFD

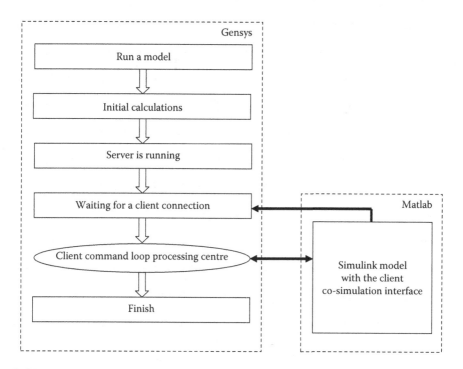

FIGURE 8.9 The flow chart for the co-simulation process in Gensys and MATLAB.

and 7FFFFFFFFFFFFFFE. In the case of the first error code, the simulation process can still continue, but for the last two error codes the process stops immediately.

Based on these commands, the co-simulation process for Gensys should be as depicted in Figure 8.9. In addition, it is necessary to mention that additional commands such as 'defgroup *<group_name>* *<no of address>* *<var_addresses>*', 'getgroup *<group_name>*' and 'putgroup *<group_name>* *<values>*' can be used for a group data exchange. This can be a very useful solution for real-time simulations [29].

8.3.2 DESIGN OF THE CO-SIMULATION INTERFACE IN SIMULINK

The client interface version (referred to in the following text as 'Client') presented in this section has been developed for MATLAB/Simulink 2012a running in a Linux environment. The basic algorithm of the version for a Windows platform can be found in [30]. For the development of the basic algorithm, it is necessary to initially develop the sequence of commands for the communication between the Client and the Server. The flowchart of such a process based on the available commands from Gensys, described in the previous section, is shown in Figure 8.10.

The S-function mechanism is a very powerful tool for writing program code and it makes the development process easy [30,31]. The S-function is written in C language for the Client and also contains the Client user interface for the input of co-simulation parameters (see Stage 1 in Figure 8.10) required for communication and data exchange between Gensys and Simulink. As Stages 2 and 3 in Figure 8.10

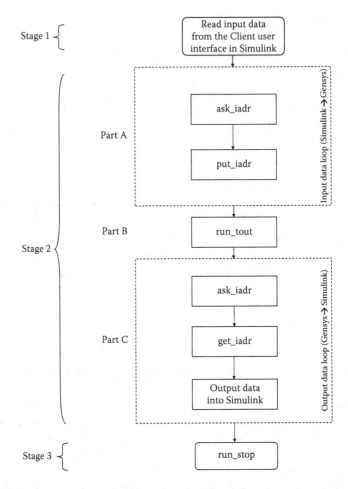

FIGURE 8.10 The command flow chart for the co-simulation process.

require writing a code, it is first necessary to open an S-Function Template which can be obtained from the Simulink Library: S-Function examples: C-file S-functions: Detailed C-MEX Template. The opened file then needs to be saved as 'Client.c'. The next step is to create an empty Simulink model file and insert the S-Function block from the Simulink Library/User-Defined Functions. Then, the S-function name in the Function Block Parameters should be defined as Client (see Figure 8.11). Stage 1 requires the development of a Client user interface. For this purpose, it is necessary to Create Mask for the S-function block and then to insert and edit the required parameters. Finally, the Client user interface can be defined as shown in Figure 8.12.

It is now time to insert a code into Client.c. For the Client, only the following S-function methods are required:

- Model initialisation stage (Stage 1 in Figure 8.10):
  - mdlStart;

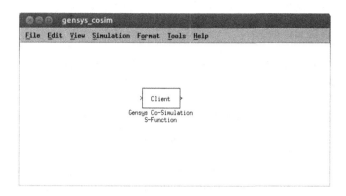

FIGURE 8.11 The S-function block in Simulink model.

- mdlCheckParameters;
- mdlInitializeSizes;
- mdlInitializeSampleTimes;
- mdlInitializeConditions.
- Simulation loop stage (Stage 2 in Figure 8.10):
  - mdlOutputs;
  - mdlTerminate.

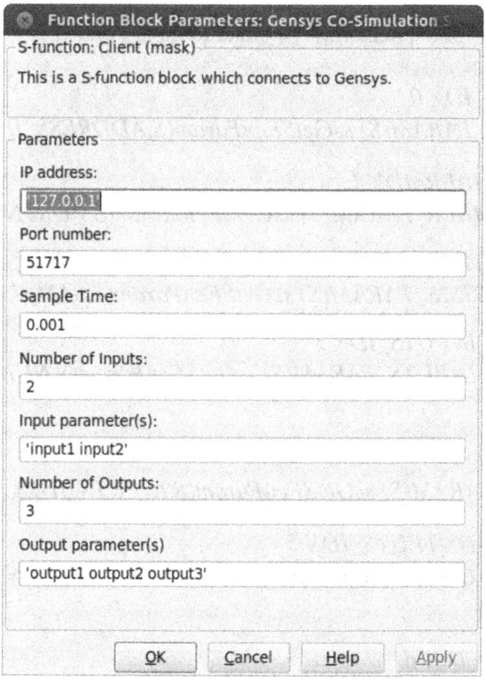

FIGURE 8.12 The Client user interface for the S-function block.

- Simulation stop (Stage 3 in Figure 8.10):
  - mdlTerminate.

At the beginning of the C-MEX Template file, it is necessary to specify the name of the S-function as:

#define S_FUNCTION_NAME Client

After that, the following header files should be added:

#include <stdio.h >
#include <stdlib.h >
#include <string.h >
#include <strings.h >
#include <arpa/inet.h >
#include <sys/types.h >
#include <sys/socket.h >
#include <netinet/in.h >
#include <netdb.h >
#include <unistd.h >
#include "mex.h"

Then, the macro substitutions are defined as

#define U(element) (*uPtrs[element]) /* Pointer to Input Port0 */

#define ADDRESS_IDX 0
#define ADDRESS_PARAM(S) ssGetSFcnParam(S,ADDRESS_IDX)

#define PORTNUMBER_IDX 1
#define PORTNUMBER_PARAM(S) ssGetSFcnParam(S,PORTNUMBER_IDX)

#define SAMPLE_TIME_IDX 2
#define SAMPLE_TIME_PARAM(S) ssGetSFcnParam(S,SAMPLE_TIME_IDX)

#define NUMBER_INPUTS_IDX 3
#define NUMBER_INPUTS_PARAM(S) ssGetSFcnParam(S,NUMBER_INPUTS_IDX)

#define INPUTS_IDX 4
#define INPUTS_PARAM(S) ssGetSFcnParam(S,INPUTS_IDX)

#define NUMBER_OUTPUTS_IDX 5
#define NUMBER_OUTPUTS_PARAM(S) ssGetSFcnParam(S,NUMBER_OUTPUTS_IDX)

#define OUTPUTS_IDX 6
#define OUTPUTS_PARAM(S) ssGetSFcnParam(S,OUTPUTS_IDX)

#define NPARAMS 7
#define MAX_NUMBER_OF_INPUTS 100

Next, the declaration list of variables can be initiated as follows:

int sockfd;
long double f;
unsigned char p[128];
int number_inputs, number_outputs;
char*input_5,*input_7;
static char*msg[35];
unsigned char*input_names[100],*output_names[100];
unsigned char server_command[128], server_output[128];
unsigned char c;
unsigned char dataparameters[100];
int numberofinputparameters, numberofoutputparameters;
char*hostname;

The 'mdlStart' function reads the names of input and output parameters in arrays for further use by other methods.

static void mdlStart(SimStruct *S)

{
*   int i;*
*   int m,l,h,g;*
*   int numberofchars;*
*   int nx;*

/* INPUT_NAMES: this section accesses the names of input parameters and stores them in the array*/

*   number_inputs=(int) mxGetScalar(NUMBER_INPUTS_PARAM(S));*
*   numberofchars=mxGetNumberOfElements(INPUTS_PARAM(S))+1;*
*   input_5=calloc(1000, sizeof(char));*
*   mxGetString(INPUTS_PARAM(S), input_5, numberofchars);*
*   for (i=0;i < MAX_NUMBER_OF_INPUTS;i++) {*
*   input_names[i]=calloc(100, sizeof(char));*
}

*   m=0; l=0;*
*   for (i = 0;i < numberofchars;i++) {*
*   c = input_5[i];*
*   if ((c==' ')||(c = ='\0')) {*
*   dataparameters[l] = '\0';*

Design and Simulation of Rail Vehicles

```
        strcat(input_names[m],dataparameters);
        l = 0;
        m++;
    }
    else {
        dataparameters[l] = c;
        l++;
    }
}

numberofinputparameters = m;

/* INPUT_NAMES: the end of this section*/
/* OUTPUT_NAMES: this section accesses the names of input parameters and
stores them in the array*/

    number_outputs = (int) mxGetScalar(NUMBER_OUTPUTS_PARAM(S));
    numberofchars = mxGetNumberOfElements(OUTPUTS_PARAM(S))+1;
    input_7 = calloc(1000, sizeof(char));
    mxGetString(OUTPUTS_PARAM(S), input_7, numberofchars);
    for (i = 0;i < MAX_NUMBER_OF_INPUTS;i++) {
        output_names[i] = calloc(100, sizeof(char));
    }

    m = 0; l = 0;
    for (i = 0;i < numberofchars;i++) {
        c = input_7[i];
        if ((c==' ')||(c=='\0')) {
            dataparameters[l] = '\0';
            strcat(output_names[m],dataparameters);
            l = 0;
            m++;
        }
        else {
            dataparameters[l]=c;
            l++;
        }
    }

numberofoutputparameters = m;

/* OUTPUT_NAMES: the end of this section*/

/*Get host name from the input data*/
    hostname = calloc(1000, sizeof(char));
    nx = mxGetNumberOfElements(ADDRESS_PARAM(S)) + 1;
```

```
mxGetString(ADDRESS_PARAM(S), hostname, nx);
/*End of this section*/
}
```

The 'mdlCheckParameters' function processes the validation and verification of parameters in the Client user interface. It can be defined as follows:

```
static void mdlCheckParameters(SimStruct *S)
{
    mwSize buflen;
/* Check 1st & 2nd parameters: ADDRESS/PORTNUMBER parameters */
    {
        if (mxGetNumberOfElements(PORTNUMBER_PARAM(S)) <= 0) {
        ssSetErrorStatus(S, "The port number (parameter) is wrong");
        return;
    }
        if ((mxGetScalar(PORTNUMBER_PARAM(S)) <= 0) ||
(mxGetScalar(PORTNUMBER_PARAM(S)) >= 65535)){
            ssSetErrorStatus(S, "The port number (parameter) is wrong");
            return;
        }
    }
......
}
```

The 'mdlInitializeSizes' function is used to set the number of inputs and outputs for the S-function. It can be defined as follows:

```
static void mdlInitializeSizes(SimStruct *S)
{
    ssSetNumSFcnParams(S, NPARAMS); /* Number of expected parameters */
if (ssGetNumSFcnParams(S) == ssGetSFcnParamsCount(S)) {
    mdlCheckParameters(S);
    if (ssGetErrorStatus(S) ! = NULL) {
        return;
    }
}
else {
    return;/* Parameter mismatch will be reported by Simulink */
}

if (!ssSetNumInputPorts(S, 1)) return;
    ssSetInputPortWidth(S, 0, number_inputs);
    ssSetInputPortDirectFeedThrough(S, 0, 1);
    ssSetInputPortOverWritable(S, 0, 1);
    ssSetInputPortOptimOpts(S, 0, SS_REUSABLE_AND_LOCAL);
```

```
if (!ssSetNumOutputPorts(S, 1)) return;
    ssSetOutputPortWidth(S, 0, number_outputs);
    ssSetOutputPortOptimOpts(S, 0, SS_REUSABLE_AND_LOCAL);

ssSetNumSampleTimes(S, 1);
ssSetNumRWork(S, 0);
ssSetNumIWork(S, 0);
ssSetNumPWork(S, 0);
ssSetNumModes(S, 0);
ssSetNumNonsampledZCs(S, 0);

ssSetOptions(S, SS_OPTION_WORKS_WITH_CODE_REUSE |
SS_OPTION_EXCEPTION_FREE_CODE |
SS_OPTION_USE_TLC_WITH_ACCELERATOR);
}
```

The value of sample time for our S-function is specified with the 'mdlInitializeSampleTimes' function as follows:

```
static void mdlInitializeSampleTimes(SimStruct *S)
{
    ssSetSampleTime(S,0,mxGetScalar(ssGetSFcnParam(S,2)));
}
```

For the establishment of the TCP/IP communication process, the 'mdlInitializeConditions' function is used as follows:

```
static void mdlInitializeConditions(SimStruct *S)
{
    int portno, n;
    struct sockaddr_in serv_addr;
    struct hostent *server;

    char buffer[256];

    /*Get port from the input data*/
    portno = (int) mxGetScalar(PORTNUMBER_PARAM(S));

    sockfd = socket(AF_INET, SOCK_STREAM, 0);
    if (sockfd < 0) {
        ssSetErrorStatus(S,"ERROR opening socket");
        }
    server = gethostbyname(hostname);
    if (server = =NULL) {
        ssSetErrorStatus(S,"ERROR, no such host");
        }

    bzero((char *) &serv_addr, sizeof(serv_addr));
    serv_addr.sin_family = AF_INET;
    serv_addr.sin_port = htons(portno);
```

```
bcopy((char *)server->h_addr, (char *)&serv_addr.sin_addr.s_addr,server->h_
length);
n = connect(sockfd, (struct sockaddr*)&serv_addr, sizeof(serv_addr));
if (n < 0) {
    ssSetErrorStatus(S, "ERROR connecting");
    }
}
```

The 'mdlOutputs' function provides data exchange between the Client and Server and sends the command 'run_tout' to execute a simulation by the Server at each time step as shown here:

```
static void mdlOutputs(SimStruct *S, int_T tid)
{
    real_T x0,x1, data;
    int_T portWidth = ssGetInputPortWidth(S,0);
    InputRealPtrsType uPtrs = ssGetInputPortRealSignalPtrs(S,0);
    real_T *y = ssGetOutputPortSignal(S,0);
    int_T i, n, ins;
        char_T command1[128];
        char_T command2[128];
    UNUSED_ARG(tid);/* not used in single tasking mode */

/* See Part A in Figure 8.10 */
/* Start sending input parameters to the server*/
    for (i = 0;i < numberofinputparameters;i + +) {
        sprintf(server_command, "ask_iadr");
        n = write(sockfd, &server_command, (strlen(server_command) + 1));
        if (n < 0) {
            ssSetErrorStatus(S, "ERROR writing to socket");
            }
        sleep(0.0001);
        n = read(sockfd, &server_command, 128);
        if (n < 0) {
ssSetErrorStatus(S, "ERROR reading from socket");
        }
    sleep(0.0001);
/********************* ask_iadr *****************************/
    n = write(sockfd, input_names[i], (strlen(input_names[i]) + 1));
        if (n < 0) {
            ssSetErrorStatus(S, "ERROR writing to socket");
            }
        sleep(0.0001);
    server_command[0] = '\0';
    n = read(sockfd, &server_command, 128);
```

```
              if (n < 0) {
                  ssSetErrorStatus(S, "ERROR reading from socket");
      }
      sleep(0.0001);
/*********************** put_iadr ****************************/
      sprintf(command1, "put_iadr");
      n = write(sockfd, &command1, (strlen(command1) + 1));
      if (n < 0) {
                  ssSetErrorStatus(S, "ERROR writing to socket");
              }
              sleep(0.0001);
      n = read(sockfd, &command1, 128);
         if (n < 0) {
                  ssSetErrorStatus(S, "ERROR reading from socket");
              }
      sleep(0.0001);
/*xxxxxxx put_iadr:first argument xxxxxx*/
      ins = atoi(server_command);
      sprintf(command1, "%d",ins);
      n = write(sockfd, &command1, (strlen(command1)) + 1);
      if (n < 0) {
                  ssSetErrorStatus(S, "ERROR writing to socket");
              }
      sleep(0.0001);
      n = read(sockfd, &server_command, 128);
      if (n < 0) {
           ssSetErrorStatus(S, "ERROR reading from socket");
          }
      sleep(0.0001);
/*xxxxxxx put_iadr:second argument xxxxxx*/
      data = ((real_T)(*uPtrs[i]));
      sprintf(command2, "%e",data);
         n = write(sockfd, &command2, (strlen(command2) + 1));
         if (n < 0) {
                     ssSetErrorStatus(S, "ERROR writing to socket");
                 }
         sleep(0.0001);
      n = read(sockfd, &server_command, 128);
         if (n < 0) {
                     ssSetErrorStatus(S, "ERROR reading from socket");
                 }
         sleep(0.0001);
    }
/* End sending input parameters to the server*/

/* See Part B in Figure 8.10 */
```

```
/* To send a command to start simulation the server */
    sprintf(server_command, "run_tout");
    n = write(sockfd, &server_command, 9);
    if (n < 0) {
            ssSetErrorStatus(S, "ERROR writing to socket");
            }
    sleep(0.0001);
    n = read(sockfd, &server_command, 128);
    if (n < 0) {
            ssSetErrorStatus(S, "ERROR reading from socket");
            }
    sleep(0.0001);
/*End of this section*/

/* See Part C in Figure 8.10 */

/* Start receiving output parameters to the server*/
    for (i = 0;i < numberofoutputparameters;i + +) {
    sprintf(server_command, "ask_iadr");
    n = write(sockfd, &server_command, (strlen(server_command) + 1));
    if (n < 0) {
            ssSetErrorStatus(S, "ERROR writing to socket");
            }
    sleep(0.0001);
    n = read(sockfd, &server_command, 128);
    if (n < 0) {
            ssSetErrorStatus(S, "ERROR reading from socket");
            }
    sleep(0.0001);
/********************* ask_iadr **********************************/
    n = write(sockfd, output_names[i], (strlen(output_names[i]) + 1));
    if (n < 0) {
            ssSetErrorStatus(S, "ERROR writing to socket");
            }
    sleep(0.0001);
    server_command[0] = '\0';
    n = read(sockfd, &server_command, 128);
    if (n < 0) {
            ssSetErrorStatus(S, "ERROR reading from socket");
            }
    sleep(0.0001);
/********************* get_iadr **********************************/
    sprintf(command1, "get_iadr");
    n = write(sockfd, &command1, (strlen(command1) + 1));
    if (n < 0) {
        ssSetErrorStatus(S, "ERROR writing to socket");
        }
```

```
         sleep(0.0001);
      n = read(sockfd, &command1, 128);
         if (n < 0) {
                    ssSetErrorStatus(S, "ERROR reading from socket");
                    }
         sleep(0.0001);
/*xxxxxxx get_iadr:first argument xxxxxx*/
   ins = atoi(server_command);
   sprintf(command1, "%d",ins);
      n = write(sockfd, &command1, (strlen(command1)) + 1);
         if (n < 0) {
                    ssSetErrorStatus(S, "ERROR writing to socket");
                    }
      sleep(0.0001);
   n = read(sockfd, &server_output, 128);
         if (n < 0) {
                    ssSetErrorStatus(S, "ERROR reading from socket");
                    }
      sleep(0.0001);
   y[i] = atof(server_output);
}
/* End receiving output parameters to the server*/
/* End of data exchange*/
}
```

The function 'mdlTerminate' is a mandatory method used to stop simulation and to close the TCP/IP connection. It is necessary to define this method as follows:

```
static void mdlTerminate(SimStruct *S)
{
   int n;
      sprintf(server_command, "run_stop");
      n = write(sockfd, &server_command, 9);
         if (n < 0) {
                    ssSetErrorStatus(S, "ERROR writing to socket");
                    }
      sleep(0.0001);
      close(sockfd);
   UNUSED_ARG(S);/* unused input argument */
}
```

After the file Client.c has been saved, it is necessary to compile it in MATLAB as follows:

```
≫mex Client.c
```

Finally, the file Client.mexglx or the file Client.mexa64 should be created (which form it takes is dependent on the architecture of the operating system, these being for 32- or 64-bits, respectively). The Client now is ready for the co-simulation.

8.3.3 Example of Co-Simulation Implementation between Gensys and Simulink

In this example, the developed client interface in MATLAB/Simulink has been used for co-simulation of the traction control system and its connection to a Gensys multi-body model. This interface has been compiled for MATLAB 2012a (64-bit version) and both products have been running under OS Ubuntu 12.04 (64 bit). The models of a hypothetical suburban train power car and its traction control system have been generated in Gensys and Simulink, respectively.

The characteristics of the vehicle model of a power car are presented in Table 8.1. In addition, the general Gensys model view of a power car is shown in Figure 8.13. The gross mass of the power car is 67.7 metric tons. The power car is equipped with two bogies with four traction axles (Bo-Bo).

TABLE 8.1
Common Parameters for the Model of the Power Car

Parameter	Value
Wheel spacing	2.38 m
Bogie spacing	14.68 m
Vehicle body mass	47340 kg
Bogie frame mass (the traction motor mass shared between bogie and axles)	6390 kg
Axle mass (the traction motor mass shared between bogie and axles)	1895 kg
Primary Suspension	
Vertical stiffness	1267 kN/m
Vertical damper	60 kN·s/m
Longitudinal stiffness	20000 kN/m
Lateral stiffness	20000 kN/m
Secondary Suspension	
Vertical stiffness	350 kN/m
Vertical damper	17 kN·s/m
Longitudinal stiffness	250 kN/m
Lateral stiffness	250 kN/m
Lateral damper	30 kN·s/m
Vertical viscous damper (series stiffness)	3500 kN/m
Vertical viscous damper	17 kN·s/m
Traction rod longitudinal stiffness	15000 kN/m
Traction rod longitudinal damper	100 kN·s/m

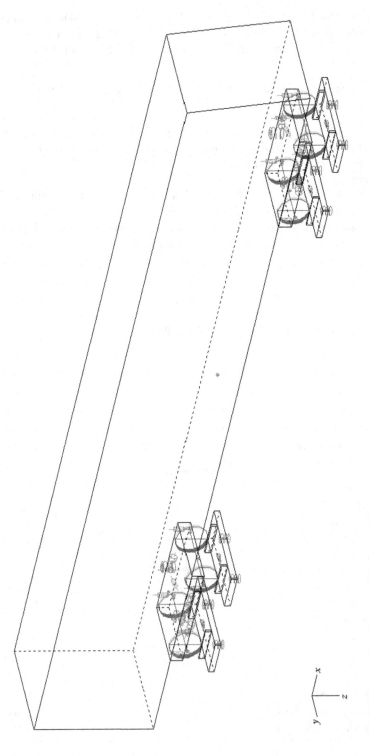

FIGURE 8.13 The Gensys multibody model of a power car.

The power car has been modelled for a standard track gauge of 1435 mm. The model includes a car body, two bogies and four axles (wheelsets), which are modelled as rigid bodies. In addition, all of them have six degrees of freedom each. The track is modelled as lumped masses under the wheels, which join with rails through their contact zones. Each track mass has three degrees of freedom (roll rotation and lateral and vertical translations).

New ANZR1 wheel and AS60 rail profiles have been used. In the wheel-rail contact modelling, a three-point contact approach is used. The contact between wheel and rail is modelled with three spring and damper elements, which are aligned normal to their three respective wheel–rail contact surfaces. The creep forces are calculated with the contact model based on the Kalker theory.

The simplified traction system has been developed for a power car that is equipped with one inverter per wheelset, that is, an individual wheelset traction control strategy is implemented in this case. The system uses a feedback control strategy as shown in Figure 8.14, where V is the linear car speed; ω is the wheelset angular velocity; s_{est} is the estimated slip; s_{opt} is the optimal slip; T_{ref} is the reference torque; T_{ref}^* is the reference torque generated by the control system; T_{in} is the input motor torque; T_{wheels} is the traction torque applied to is the wheelset and ΔT is the torque reduction. The inverter and traction motor dynamics can be written as

$$T_{wheels} = \frac{1}{\tau s + 1} T_{in} \cdot i \tag{8.1}$$

where τ is a time constant and i is a gear ratio.

The slip observer is implemented in Gensys (see Figure 8.14) and the estimated slip for each axle (wheelset) is calculated as

$$s_{est} = \frac{\omega r - V}{\omega r} \tag{8.2}$$

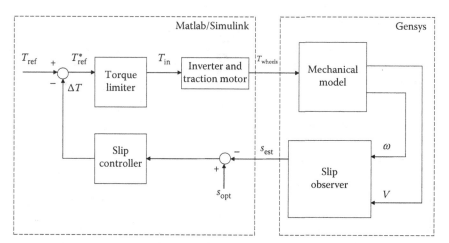

FIGURE 8.14   The traction control system of a power car for one axle.

where r is the wheel radius. The torque limiter does not allow the control system to exceed the reference torque. The slip controller is a simple P controller, which uses a slip error as the input signal to the controller.

The modelling has been done by means of co-simulation for the constant linear speed of 30 km/h for the power car. The adjustable longitudinal coupler force, dependent on the traction torques, has been attached to the model in order to provide a realistic simulation condition. The car is running on a straight track with no track alignment errors.

The simulation has been performed with the following assumptions regarding the power car's design and tractive effort characteristics:

- The maximum available adhesion is 0.14;
- The optimal slip is 0.04 (this means the stable zone for maximum adhesion is between slip values of 0.03 and 0.05);
- The reference torque value is 3000 N m and the limiter torque values are ±3500N;
- The gear ratio is 3.15.

The integration approach is based on the aggregation of relevant sub-models into one model (Figure 8.14), which has been described above. The traction control system designed in Simulink is shown in Figure 8.15. The system has 4 sub-systems

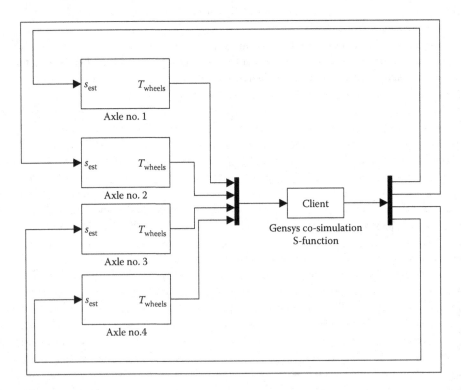

FIGURE 8.15 The traction control system of a power car in Simulink.

T_{ref}
T_{ref}
S_{opt}
S_{opt}
1
S_{est}
K_p
Proportional gain
Torque limiter
zoh 1
$t{\cdot}s+1$
Transfer fcn
Scope 4
i
Gear ratio
1
T_{wheels}

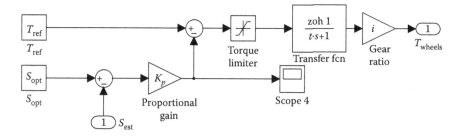

FIGURE 8.16 The traction control subsystem for one axle.

(each one for one axle) and has 4 inputs (providing the estimated slip values, s_{est}, into the traction control system) and 4 outputs (sending the torque values, T_{wheels}, into a mechanical system). The traction control sub-system for one axle is shown in Figure 8.16. The Client user interface settings are shown in Figure 8.17.

During the simulation, a modified Heun's method with a time step of 1 ms has been used as the Gensys solver. In Simulink, the discrete (no continuous states) solver with a fixed-step size of 1 ms has been chosen.

The simulation results with the constant speed, presented in Figure 8.18, show that the system works properly and the desired slip and adhesion values between

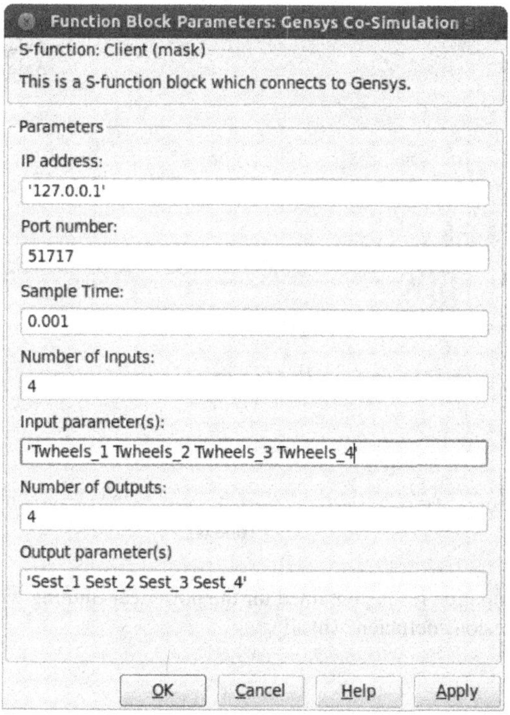

Function Block Parameters: Gensys Co-Simulation
S-function: Client (mask)
This is a S-function block which connects to Gensys.
Parameters
IP address:
'127.0.0.1'
Port number:
51717
Sample Time:
0.001
Number of Inputs:
4
Input parameter(s):
'Twheels_1 Twheels_2 Twheels_3 Twheels_4'
Number of Outputs:
4
Output parameter(s)
'Sest_1 Sest_2 Sest_3 Sest_4'
OK
Cancel
Help
Apply

FIGURE 8.17 The Client user interface settings.

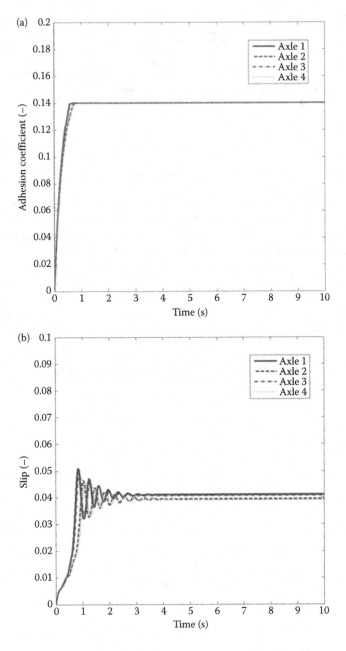

FIGURE 8.18 Simulation results obtained for the power car running with a constant speed of 40 km/h. (a) Adhesion coefficient, (b) slip.

wheel and rails have been reached. Some oscillation of the slip values can be explained by the fact that the simplified proportional slip controller has been used. In addition, the difference in slip results appears to be due to the dynamic weight distribution between axles.

Finally, the simulation results mean that the developed co-simulation interface works and the goal of this chapter has been achieved.

REFERENCES

1. W. Körtum, O. Vaculín, Is multibody simulation software suitable for mechatronic systems? *Proceedings of the 5th World Congress on Computational Mechanics*, Vienna, Austria, 7–12 July, 2012.
2. M. Arnold, A. Carrarini, A. Heckman, G. Hippmann, Simulation techniques for multidisciplinary problems in vehicle system dynamics, in *Computational Mechanics in Vehicle System Dynamics, Supplement to Vehicle System Dynamics*, M. Valášek (Ed.), Taylor & Francis, London, UK, 2004, Vol. 40, pp. 17–36.
3. M. Arnold, B. Burgermeister, C. Führer, G. Hippmann, G. Rill, Numerical methods in vehicle system dynamics: State of the art and current developments, *Vehicle System Dynamics*, 49(7), 2011, 1159–1207.
4. MODELISAR, Functional mock-up interface for co-simulation, Document version 1.0, 12 October, 2010.
5. MODELISAR, Functional mock-up interface for model exchange and co-simulation, Document version 2.0 Beta 4, 10 August, 2012.
6. A. Chudzikiewicz, Simulation of rail vehicle dynamics in MATLAB environment, *Vehicle System Dynamics*, 33(2), 2000, 107–119.
7. E. Meli, M. Malvezzi, S. Papini, L. Pugi, M. Rinchi, A. Rindi, A railway vehicle multibody model for real-time applications, *Vehicle System Dynamics*, 46(12), 2008, 1083–1105.
8. INTEC GmbH, SIMAT—The link between SIMPACK and MATLAB, Simpack Release 8.9, 11 October, 2008.
9. INTEC GmbH, SIMPACK Code Export, Simpack Release 8.9, 11 October, 2008.
10. INTEC GmbH, SIMPACK MatSIM, Simpack Release 8.8, 29 October, 2006.
11. A.C. Zolotas, J.T. Pearson, R.M. Goodall, Modelling requirements for the design of active stability control strategies for a high speed bogie, *Multibody System Dynamics*, 15(1), 2006, 51–66.
12. M. Spiryagin, K.S. Lee, H.H. Yoo, Control system for maximum use of adhesive forces of a railway vehicle in a tractive mode, *Mechanical Systems and Signal Processing*, 22(3), 2008, 709–720.
13. M. Spiryagin, V. Spiryagin, Modelling of mechatronic systems of running gears for a rail vehicle, East Ukrainian National University, Lugansk, Ukraine, 2010. ISBN: 9789665908715 (in Ukrainian).
14. N. Bosso, A. Gugliotta, A. Somà, M. Spiryagin, Model of scaled test rig for real time applications, *Proceedings of the 21st International Congress of Mechanical Engineering*, Natal, Brazil, 24–28 October, 2011.
15. MathWorks, *Code Generation from MATLAB User's Guide, Release* 2012b. See: http://www.mathworks.com/help/pdf_doc/eml/eml_ug.pdf.
16. MSC Software, Welcome to Adams/Controls, Adams MD R3, 9 October, 2009. See: http://simcompanion.mscsoftware.com/infocenter/index?page=content&id=DOC9300&cat=11P7155&actp=LIST.

17. MSC Software, Getting started using Adams/Controls, 13 January, 2012. See: http:// simcompanion.mscsoftware.com/infocenter/index?page=content&id=DOC10081& cat=1VMO50&actp=LIST.
18. S. Müller, R. Kögel, R. Schreiber, Numerical simulation of the drive performance of a locomotive on straight track and in curves, *14th European ADAMS Users' Conference*, Berlin, Germany, 1999.
19. S. Müller, R. Kögel, R. Schreiber, Simulation of a locomotive as a mechatronical system, *4th ADAMS/Rail Users Conference*, Utrecht, The Netherlands, 1999.
20. O. Polach, Influence of locomotive tractive effort on the forces between wheel and rail, *Vehicle System Dynamics*, 35(Suppl), 2001, 7–22.
21. J.C. Huang, J. Xiao, H. Weiss, Simulation study on adhesion control of electric locomotives based on multidisciplinary virtual prototyping, *Proceedings of the IEEE International Conference on Industrial Technology (ICIT 2008)*, Chengdu, China, 21–24 April, 2008.
22. Vampire User Guide—Part S, Simulink Interface, Version 4.32, pp. 453–499.
23. Universal Mechanism, User's manual—UM co-simulation tool—Version 6.0, 2010.
24. Universal Mechanism, Interface with MATLAB/Simulink—Version 6.0, 2010.
25. R.V. Kovalev, G.A. Fedyaeva, V.N. Fedyaev, *Modelling of an electro-mechanical system of locomotives*, Sbornik Trudov of DIIT, no. 14, 2007, pp. 123–127 (in Russian). See: http://www.universalmechanism.com/index/download/elmechloco.pdf.
26. V.N. Fedyaev, Influence of electrical and mechanical subsystems of a shunting locomotive on the realization limit tractive efforts, PhD thesis, Bryansk State Technical University, Bryansk, Russia, 2006 (in Russian).
27. G.A. Fedyaeva, Forecasting of dynamic process under transient and emergency mode for traction drives with asynchronous motors, DSc thesis, Moscow State Railway University, Moscow, Russia, 2008 (in Russian).
28. AB DEsolver, *GENSYS.1309 Reference Manual—Users Manual for Program CALC*. See: http://gensys.se/doc_html/calc.html.
29. M. Spiryagin, Y.Q. Sun, C. Cole, T. McSweeney, S. Simson, I. Persson, Development of a real-time bogie test rig model based on railway specialised multibody software, *Vehicle System Dynamics*, 51(2), 2013, 236–250.
30. M. Spiryagin, S. Simson, C. Cole, I. Persson, Co-simulation of a mechatronic system using Gensys and Simulink, *Vehicle System Dynamics*, 50(3), 2012, 495–507.
31. M. Sysel, TCP/IP output from the Simulink, *Proceedings of the 17th International Conference on Process Control*, Štrbské Pleso, Slovakia, 9–12 June, 2009, pp. 634–637.

9 Advanced Simulation Methodologies

The simulation of rail vehicles requires the implementation of multidisciplinary knowledge, and it is clear that various evaluation methodologies cannot be performed using the multibody software products without the consideration of information and data from other scientific areas and/or experiments. The common situations (how and when to use advanced simulation?) are examined and discussed in this chapter.

9.1 COMPLEX TASKS AND THEIR SOLUTIONS

From previous chapters, it is possible to see that the rail vehicle is a very complex system and detailed modelling requires the development of many subsystems and the use of a diverse array of input parameters. The development of subsystems can be covered by the application of the co-simulation technique as described in Chapter 8. However, how is it possible to deal with uncertainties in input parameters? Relevant examples include friction characteristics between wheels and rails or between brake shoes and wheels, or forces from the dynamic action of the train which should be used in multibody models and so on. Researchers commonly use four standard approaches to compensate for a lack of the required data:

- Engineering analysis;
- Additional simulations in other software products;
- Laboratory testing;
- Field tests.

Engineering analysis is based on the study of existing publications and patents in the appropriate technical areas in order to retrieve some relevant information on which to base the assumptions necessary to produce required data for further multibody simulations.

Additional simulations are required when we need to retrieve some data from other areas, for example, finite element analysis for the development of multibody models with flexible bodies instead of commonly used rigid ones.

Laboratory testing is required to get more accurate or detailed information about characteristics of components included in the model. Examples include friction damper or draft gear behaviours, which have strongly non-linear characteristics and for which an assumption of linearisation is not acceptable to achieve dependable model accuracy.

Field testing is commonly required by railway regulators when the train/track interaction behaviour is to be taken into consideration. However, this approach is an increasingly expensive one and it is less often used nowadays.

In some cases, it is necessary to ask what can be done when published information is limited or other ways of obtaining it are unavailable. In such circumstances, researchers or engineers need to make realistic and justifiable assumptions for this work and declare them clearly in their reports, documentation or publications in order to avoid misinterpretation of results obtained during simulations.

The examples in the following sections can be useful for people who are going to do advanced simulation such as of a rail vehicle when train dynamics are involved in the process, or when creep forces are involved for the study of traction or braking, or multibody development for single-component testing in a real-world laboratory environment.

As in the previous chapter, Gensys software is used for the multibody simulation. In addition, longitudinal train dynamics simulations are conducted using the CRE-LTS software [1,2] in order to obtain coupler forces in cases when these were required.

9.2 SCENARIO A: ON-LINE SIMULATION AND EXISTING PRE-CALCULATED DATA

The following task commonly arises when it is necessary to investigate a wagon's dynamic performance. Let us assume that the task requires to be focussed on the study of wagon behaviour for curves of radius 500 m. The loaded wagon runs in the train on track with US Federal Railroad Administration (FRA) Class 5 track irregularities [3], which are superimposed over the designed track geometry to model typical track roughness. The track is dry with a friction coefficient of 0.4. The train speed is 20 km/h, at which the locomotives achieve their maximum tractive effort. How do we find a solution? Let us go step by step and solve this task.

9.2.1 STEP 1: DEFINITION OF SIMULATION METHODOLOGY

To evaluate the effect of lateral coupler forces on wagon dynamics, a methodology is required that includes a combination of simulations with longitudinal train dynamics and multibody packages. The proposed methodology consists of three inter-linked stages, a flowchart of which is shown in Figure 9.1.

Stage 1 considers longitudinal train dynamics, which are needed to model forces and motions of connected railway vehicles along the track direction. Commonly, as described in Chapter 6, locomotives and wagons are modelled as lumped masses, with flexure modelled using stiffness elements, connected with non-linear coupling systems that account for inbuilt coupler slack and draft gear characteristics. Locomotive traction, dynamic braking and pneumatic braking, as well as gradients, curvature and retardation/drag forces should be taken into consideration for this stage in order to get more accurate results. The main output results are coupler forces and coupler angles for selected wagons, which are required for subsequent multibody simulations.

Stage 2 covers the development of a multibody railway vehicle model, which is constructed from rigid masses connected to each other via coupling elements or

Stage 1: Longitudinal train dynamics simulation

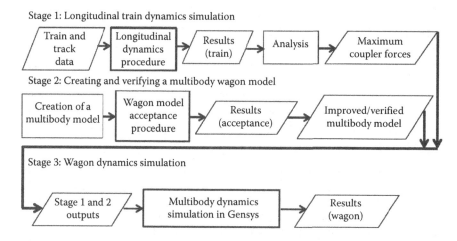

FIGURE 9.1   Simulation methodology stages.

joints in order to form a whole multibody system. Each body can have up to six degrees of freedom (longitudinal, lateral, vertical, roll, pitch and yaw) unless other restraints need to be stated in the multibody code. Commonly, for such studies, rigid bodies are used to represent the wagon car body, bogie frames and wheelsets with the dimensions, masses and moments of inertia for each being specified. Coupling elements and joints are defined on each body to connect elements such as springs, dampers, and wheel–rail contact elements. Inputs are normally given at the wheelsets [4] in response to changes in track geometry and irregularities. For the development of a valid model, several acceptance tests are required in order to identify and remedy the faults, including the tests from a Wagon Model Acceptance Procedure (WMAP) developed at the Centre for Railway Engineering [5].

Using the 'verified' wagon model developed in Stage 2, the full simulation, which includes the usage of lateral coupler forces obtained in Stage 1, should allow the achievement of obtaining a final solution in Stage 3. The lateral coupler forces are applied at both the front and rear wagon couplers at a nominal height above rail level; these forces change with respect to track geometry and longitudinal train dynamics as the train proceeds. The wagon should be tested over the required track geometry with different lateral coupler forces applied depending on different wagon positions in the train. The output data from such simulations will allow us to make some conclusions about the wagon dynamics during train operation.

9.2.2   STEP 2: SIMULATIONS

9.2.2.1   Stage 1: Longitudinal Train Dynamics Simulation

A hypothetical train set consists of 3 locomotives arranged at the front and hauling 55 identical wagons. Their masses are listed in Table 9.1. Whilst travelling over the track section, the train speed is close to 20 km/h, the maximum speed at which typical Australian AC locomotives develop their maximum continuous tractive effort [6,7].

TABLE 9.1

Train Masses (55 Wagons)

Loco	No. Locos	Loco Mass (Tonne)	Wagon	Loaded Wagon Mass (Tonne)	Loaded Train Mass (Tonne)
Type—Freight locomotive (Co–Co) with AC traction motors; Tractive effort—Continuous 450 kN at 20 km/h	3	134	Type—Mineral wagon; 2 bogies; 2 wheelsets per bogie	80	4802

A standard coupler has been placed between each pair of wagons along the train length, which is common in Australian mineral trains [8].

A hypothetical (designed) track consists of two curved sections and has a total length of 10 km. The operational data for the hypothetical track is presented in Table 9.2 and is also shown in Figure 9.2. The track has zero elevation and the track cant on the 500 m radius curves is 65 mm.

The calculation results obtained from simulation in the CRE-LTS package are shown in Figures 9.3 and 9.4. Analysing the lateral coupler force results, only one simulation case has been chosen as inputs for the further study in Stage 3, when the front and rear lateral coupler forces are approximately equal to 5 kN as shown in Figure 9.3 for wagon no. 1. In this case, maximum lateral force values at both coupler ends are taken for the evaluation of wagon behaviour in the specified curve with radius of 500 m. In the real world, many other cases should be taken for the study, but just one case is given here for simplicity.

9.2.2.2   Stage 2: Creating and Verifying a Multibody Wagon Model

A typical freight wagon operating in Australian railways has been selected for the modelling. The multibody model, shown in Figure 9.5, includes one wagon car body and two bogies. Each bogie comprises one bolster, two sideframes, four axle boxes

TABLE 9.2

Track Geometry for Longitudinal Track Dynamics Simulation

Distance from Start (m)	Track Section
0–4995	4995 m tangent track
4995–5050	55 m entry transition
5050–5595	545 m of 500 m radius curve to right
5595–5650	55 m exit transition
5650–5695	45 m tangent track
5695–5750	55 m entry transition
5750–6295	545 m of 500 m radius curve to left
6295–6350	55 m exit transition
6350–10,000	45 m tangent track

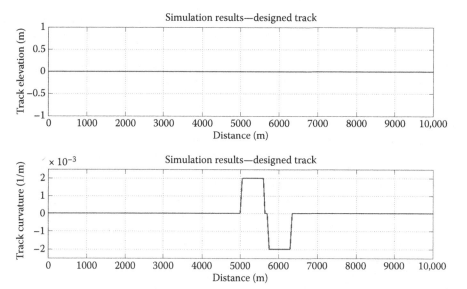

FIGURE 9.2   Hypothetical track data.

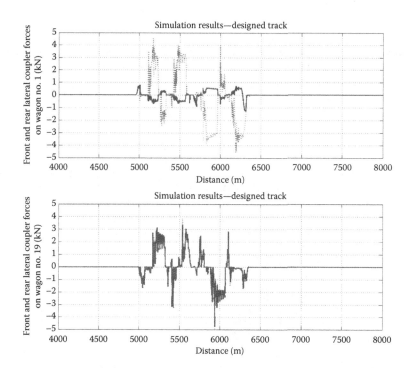

FIGURE 9.3   Example of lateral coupler force results for wagon nos. 1 and 19 in the train running on the hypothetical track (⎯⎯ front coupler; ….. rear coupler).

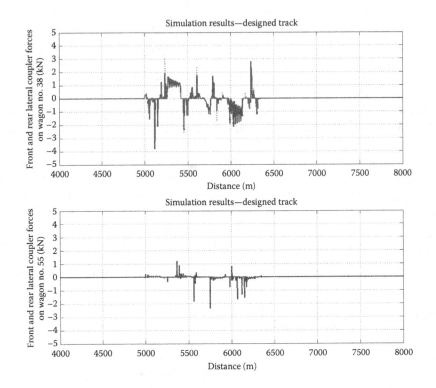

FIGURE 9.4 Example of lateral coupler force results for wagon nos. 38 and 55 in the train running on the hypothetical track (⎯⎯ front coupler; ….. rear coupler) (the lateral force on the rear coupler of wagon 55 is equal to 0).

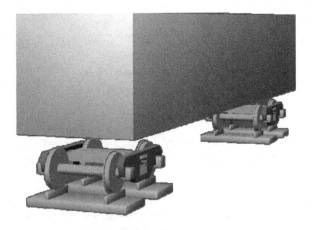

FIGURE 9.5 Multibody model in Gensys.

and two wheelsets. All these bodies are modelled as rigid mass bodies with six degrees of freedom. However, some constraints are set on these bodies as listed in Table 9.3. When the wheelset pitches (i.e. rotates as it rolls along the track), the contact point moves around the wheel. When the pitch angle is 3.14159 radians, then the contact point of the wheel will be on top of the wheel, which is not possible. Therefore, it is always necessary to put constraints on the wheelset pitch angle itself and only allow the pitch angle velocity as a degree of freedom.

The wagon modelling approach has been developed taking into account the experimental and theoretical research performed at the Centre for Railway Engineering [9–16]. The main parameters of the developed wagon model are presented in Table 9.4.

The centre bowl connection between a wagon car body and a bolster is modelled with the following elements:

• Linear stiffness element acting in all directions except yaw rotation;
• Linear damping element acting in all directions;
• Yaw series stiffness element with a moment of friction acting with yaw rotation, which is calculated for each time step.

The secondary suspension between bolster and sideframe is modelled as follows:

• The spring nest is modelled as two vertical coil-springs in parallel and two corresponding three-dimensional damping elements;
• Longitudinal constraint is provided by one longitudinal bumpstop;
• Lateral constraint is provided by one lateral bumpstop;
• Each friction wedge between the bolster and the sideframe is modelled as a massless block and the exact triangular shape is considered.

The wedge backside (slope) contact with the bolster is modelled as one contact stiffness element, which can accommodate the relative pitch rotation between bolster and sideframe, associated with one-dimensional stiffness in series with a friction block in the tangent direction. The latter is intended to present the break-out characteristic of friction between the wedge and sideframe.

TABLE 9.3
Constraints on Bodies

	x-Longitudinal	y-Lateral	z-Vertical	f-Roll	k-Pitch	p-Yaw
Wagon car body	√	√	√	√	√	√
Bolster	√	√	√	√	√	√
Sideframe	√	√	√	√	√	√
Axle box	√	√	√	√	√, $v_k = 0$	√
Wheelset	√	√	√	√	√, $k = 0$	√

Note: √–Degree considered; $k = 0$ and $v_k = 0$ refer to wheelset pitch angle and axle box angle velocity, respectively, being fixed to be equal to zero.

TABLE 9.4
Parameters of Wagon Subsystem

Parameter	Value
Car body mass (empty)	10656 kg
Mass moment of inertia of car body about x-axis	15096 kg·m^2
Mass moment of inertia of car body about y-axis	205803 kg·m^2
Mass moment of inertia of car body about z-axis	208893 kg·m^2
Car body mass (loaded)	70656 kg
Mass moment of inertia of car body about x-axis	208155 kg·m^2
Mass moment of inertia of car body about y-axis	1336811 kg·m^2
Mass moment of inertia of car body about z-axis	1354104 kg·m^2
Mass of bolster	600 kg
Mass moment of inertia of bolster about x-axis	270 kg·m^2
Mass moment of inertia of bolster about y-axis	20 kg·m^2
Mass moment of inertia of bolster about z-axis	270 kg·m^2
Mass of sideframe	400 kg
Mass moment of inertia of sideframe about x-axis	20 kg·m^2
Mass moment of inertia of sideframe about y-axis	140 kg·m^2
Mass moment of inertia of sideframe about z-axis	150 kg·m^2
Mass of axle box	50 kg
Mass moment of inertia of axle box about x-axis	0.89 kg·m^2
Mass moment of inertia of axle box about y-axis	0.78 kg·m^2
Mass moment of inertia of axle box about z-axis	0.89 kg·m^2
Mass of wheelset	1400 kg
Mass moment of inertia of wheelset about x-axis	750 kg·m^2
Mass moment of inertia of wheelset about y-axis	100 kg·m^2
Mass moment of inertia of wheelset about z-axis	750 kg·m^2

Other Dimensions	
Longitudinal distance from the mass centre of car body to the mass centre of front and rear bolsters	6.00 m
Semi-lateral distance between left and right secondary suspensions in a bogie	0.57 m
Height between the mass centre of car body (empty) and bolster	0.70 m
Height between the mass centre of car body (loaded) and bolster	2.40 m
Height between the mass centre of bolster and sideframe	0.10 m
Height between the mass centre of wheelset and sideframe	0.04 m
Semi-longitudinal distance between wheelsets in a bogie	0.90 m
Wheel radius	0.46 m

The wedge friction surface contact with the sideframe column is modelled as one stiffness element in the normal direction associated with one two-dimensional friction block at the tangent surface, and one small viscous damping element.

The connection between the wedge bottom side and the bolster is modelled as one three-dimensional stiffness element coupling with a vertical pre-load.

The primary suspension between a sideframe and an axle box is modelled as:

- One linear vertical stiffness element;
- One linear vertical damping element on each axle box;
- One longitudinal bumpstop;
- One lateral bumpstop.

Constraints between each wheelset and its axle boxes are used in the model. In the lateral direction, the linear spring element is used between an axle box and a wheelset in order to model a possible clearance between the two bodies.

In wheel–rail contact modelling, the rails are modelled as massless bodies (see Figure 9.6). There are three springs normal to the wheel–rail contact surface, so that three different contact surfaces can be in contact simultaneously. The normal contact forces are solved by these three springs. The rails are connected to the track via springs and dampers in the lateral and vertical directions. The calculations of creep forces are made in a lookup table calculated by the Fastsim algorithm developed by Kalker [17].

A simplified WMAP has been used to validate a multibody model with non-linear elements. For this purpose, only common multibody verification approaches have been involved in the verification process. The information on the performed tests is presented in Table 9.5.

The model was then checked in order to see if any numerical instability occurred during simulation. The model can be considered as a stable one with a time step of 0.001 s for the two-step Runge–Kutta method (heun_c in Gensys) with a step size control which makes back steps if the tolerance is not reached.

A further time-stepping analysis was then undertaken to determine the empty wagon's critical speed. On a 3 km length of straight, level, AS60 kg/m profile ideal track, the wagon model started off at a speed of 150 km/h, slowing down at a rate of 5 km/h/s. The normal time step of 0.001 s was used and an initial excitation was applied to the car body to induce hunting [18]. As seen in Figure 9.9, the empty

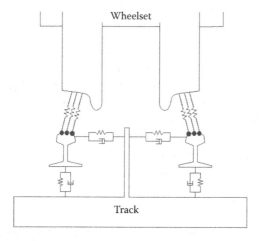

FIGURE 9.6 Track modelling in Gensys.

TABLE 9.5

Brief WMAP Tests for the Multibody Model Validation Process

Test	Status	Comments
1. Automatic syntax error checking [18–21]	Passed	
2. Visual model check [19]	Passed	See Figure 9.7
3. Time-stepping analysis—numerical instabilities [19]	Passed	Time step of 0.001 s has been chosen for further simulations
4. Critical speed [19]	Passed	See Figure 9.8

wagon stops hunting at an approximate speed of 90 km/h. This is close to 110% of its design speed (88 km/h) as described in [22], although it must be stated that this method of determining the critical speed is only accurate to within ±10%.

The obtained results show that the developed model passed the required tests, and it makes it possible to use it for the further study of wagon stability.

9.2.2.3 Stage 3: Wagon Dynamics Simulation

In Stage 3, wagon dynamics simulation is used for the assessment of running safety. Running safety of the wagon in curves can be investigated by means of the assessment of the following criteria [22]:

- The individual wheel L/V (lateral to vertical forces) ratio over a period of 0.05 s of ≤1.0;
- Sum L/V axle ratio over 0.05 s of ≤1.5;
- Wheel unloading ≤90%.

The loaded wagon travelled through a 500 m radius curve at 20 km/h with the geometry as described in Table 9.6. The applicable track cant is 65 mm for the 500 m radius curve. As specified previously, the test track has FRA class 5 irregularities (see Figure 9.9).

The approximate lateral coupler force magnitudes applied to the model, based on the results from the longitudinal train simulation tests, are shown in Figure 9.10.

The results obtained (see Figures 9.11 through 9.13) show that the fully loaded wagon has no problems during running on the track with FRA class 5 track irregularities. Not one of the running safety criteria exceeds its limit value. This means that our task is solved in this particular case, and no further investigation is required.

9.3 SCENARIO B: ON-LINE SIMULATION AND EXPERIMENTAL DATA

The following task can be applicable when it is necessary to investigate locomotive dynamics. Let us assume that the experimental curve for adhesion coefficient has been obtained for a heavy haul locomotive for the dry friction condition, as shown in Figure 9.14, and it is necessary to implement this information for the study of locomotive dynamics through a simulation process.

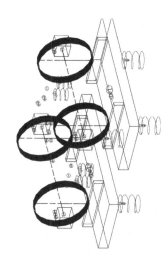

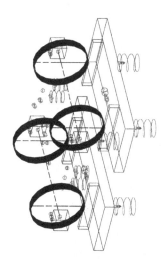

FIGURE 9.7 Visual model check in Gensys GPLOT (car body not shown in this case).

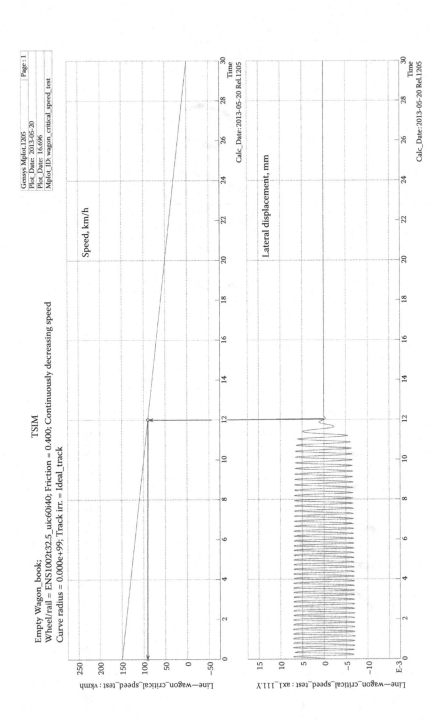

FIGURE 9.8 Non-linear stability analysis for the detection of the critical speed of the leading wheelset.

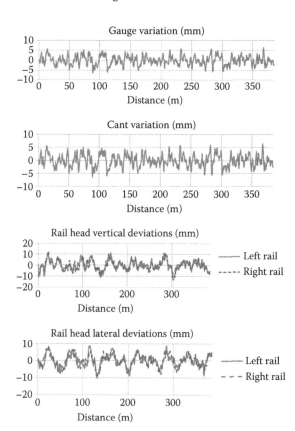

FIGURE 9.9   FRA class 5 irregularities.

Initially it is necessary to define a variable friction coefficient by means of a methodology developed by Polach [23]. This methodology has been tested for different types of locomotives and has shown very good results for the friction process characterisation in traction mode.

In Polach's model, the variable friction coefficient [24] can be expressed by

$$\mu = \mu_s((1 - A)e^{-Bw} + A) \tag{9.1}$$

TABLE 9.6
Track Geometry for Wagon Dynamics Simulation

Distance from Start (m)	Track Section
0–45	45 m tangent track
45–100	55 m entry transition
100–300	200 m right curve
300–355	55 m exit transition
355–400	45 m tangent track

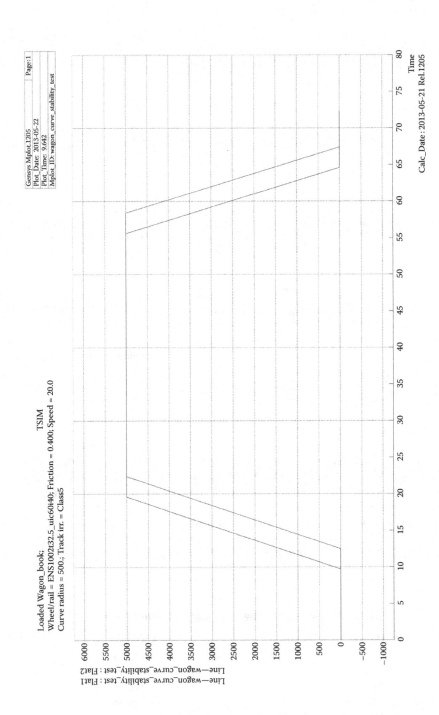

FIGURE 9.10   Approximate lateral coupler forces.

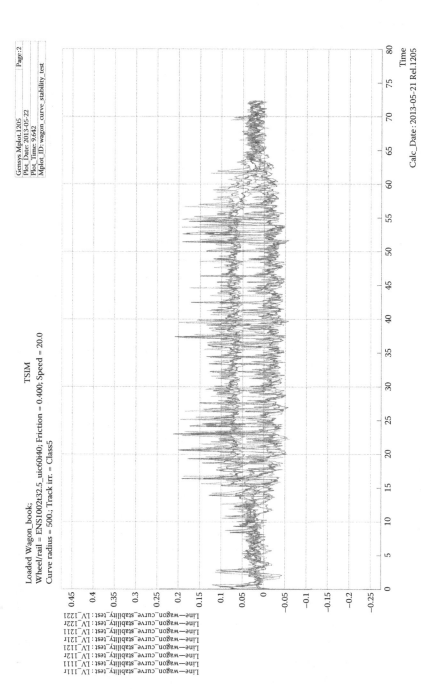

FIGURE 9.11 Running safety simulation results for the individual wheel L/V ratio.

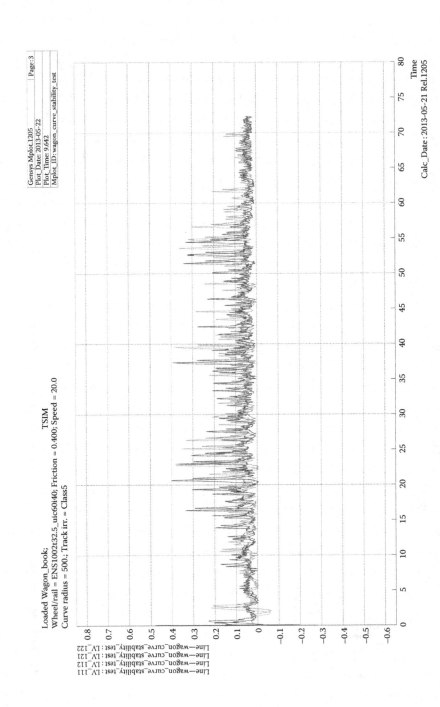

FIGURE 9.12 Running safety simulation results for axle sum L/V ratio.

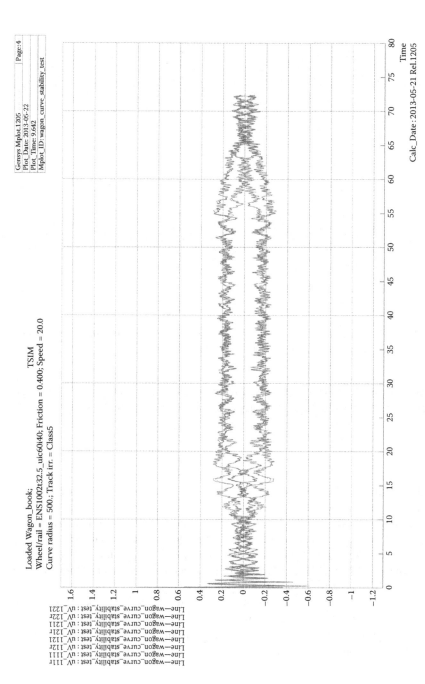

FIGURE 9.13 Running safety simulation results for wheel unloading.

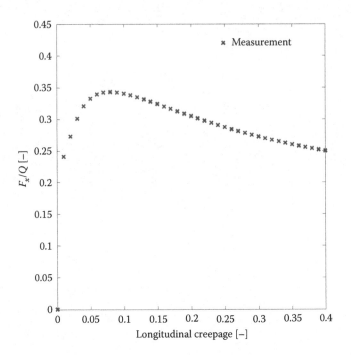

FIGURE 9.14 Adhesion versus longitudinal creepage for a locomotive running in traction mode under dry friction condition at a speed of 22 km/h.

where μ_s is the maximum coefficient of friction; A is the ratio of limit friction coefficient μ_∞ at infinity slip velocity to maximum friction coefficient μ_s, that is

$$A = \frac{\mu_\infty}{\mu_s} \tag{9.2}$$

B is the coefficient of exponential friction decrease, s/m, and w is the magnitude of the slip (creep) velocity vector.

Analysing a curve presented in Figure 9.12, some initial conditions have been obtained. After that, a set of numerical simulations have been performed in MATLAB in order to find more accurate values of coefficients in Equation 9.1. The final results of these simulations for the dry friction condition are presented in Table 9.7 and an adhesion curve is shown in Figure 9.15.

For the modelling of vehicle dynamics behaviour, the contact model used in the Gensys multibody software is based on the Polach theory [23]. Some initial parameters are required to be obtained for this model, and this can be done easily in MATLAB based on the following equations:

$$F = \frac{2Q\mu}{\pi}\left(\frac{k_A e}{1 + (k_A e)^2} + \arctan(k_S e)\right), \quad k_S \le k_A \le 1 \tag{9.3}$$

TABLE 9.7

Model Parameters for a Variable Friction Coefficient Calculation

Wheel–Rail Conditions	Dry
V (km/h)	21
μ_S	0.41
A	0.35
B (s/m)	0.45

$$e = \frac{2C\pi a^2 b}{3Q\mu} s \qquad (9.4)$$

$$s = \sqrt{s_x^2 + s_y^2} \qquad (9.5)$$

$$F_x = F\frac{s_x}{s} \qquad (9.6)$$

where k_A and k_S are model parameters for different friction conditions, a and b are the length of the semi-axes of the elliptical contact patch, s_x and s_y are the longitudinal and lateral creepages, F_x is the longitudinal creep force, Q is the wheel load and C is a proportionality coefficient characterising the contact shear stiffness.

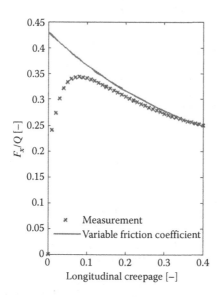

FIGURE 9.15 Adhesion and variable friction coefficients versus longitudinal creepage for dry friction condition.

TABLE 9.8

Model Parameters for Adjustment of the Creep Force Model

Freight Locomotive (Co-Co)	134 Tonne
k_A	0.6
k_S	0.13

The simulation in MATLAB has been executed for a locomotive (weight 134 tonne), which has longitudinal creepage varied from 0 to 0.4 for one wheel–rail contact point. The coefficients required for Equation 9.3 are shown in Table 9.8.

The results obtained for the wheel–rail contact are presented in Figure 9.16 and show a good agreement with measurement data.

The influence of the vehicle velocity on the form of maximum adhesion coefficient in the longitudinal direction with the proposed methodology has been modelled in MATLAB and is shown in Figure 9.17.

Now it is necessary to carry out a simulation in Gensys. The locomotive model is similar to the models published in [25,26] and has the characteristics presented in Table 9.9 and shown in Figure 9.16. The locomotive has two bogies with six traction axles, which are each equipped with an AC motor.

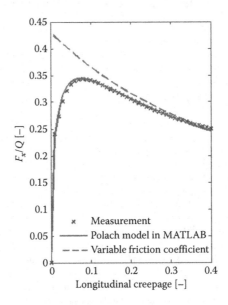

FIGURE 9.16 Results for the model with the variable friction coefficient simulated in MATLAB.

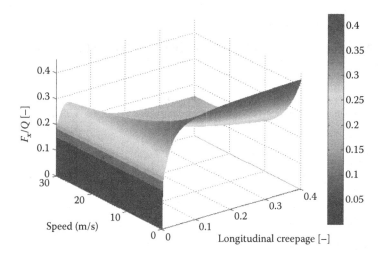

FIGURE 9.17 Calculated dependence of maximum adhesion coefficient on longitudinal creepage and vehicle velocity for the required friction condition.

The locomotive was designed with standard gauge railway bogies (1435 mm). The locomotive body, two bogies and six wheelsets are defined as rigid bodies and have six degrees of freedom each. Figure 9.18 shows the locomotive modelling for the secondary suspension:

- Six vertical coil-spring elements;
- Four lateral viscous dampers;
- Four yaw viscous dampers;
- Two lateral bumpstops;
- Four vertical bumpstops;
- Two centre pin connections.

The primary suspension between one bogie frame and three wheelsets includes the following:

- Six stiffness and six damping connections with zero length;
- At each of these connections, the longitudinal, lateral and vertical stiffness and damping coefficients are taken into account.

The wheel and rail profiles used are standard new wheel S1002 and rail UIC60 profiles. The track is modelled as lumped masses under wheelsets, and links to the wheelsets through the wheel–rail contact zones. Each track mass has three degrees of freedom (lateral and vertical translations and roll rotations). The model includes rails which are modeled as massless blocks under wheels (Figure 9.18). The rail is connected to the track via lateral and vertical springs and dampers. The track is allowed to have translations in the lateral and the vertical directions, and rotation

TABLE 9.9

Parameters for the Multibody Model of the Freight Locomotive

Parameter	Value
Vehicle Body	
Centre of gravity, vertical	1.930 m
Mass	91600 kg
Moment of inertia, roll	177095 kg·m²
Moment of inertia, pitch	3793457 kg·m²
Moment of inertia, yaw	3772695 kg·m²
Bogie Frame	
Centre of gravity, vertical	0.733 m
Mass	11000 kg
Moment of inertia, roll	4826 kg·m²
Moment of inertia, pitch	33585 kg·m²
Moment of inertia, yaw	37234 kg·m²
Axles (The Traction Motor Mass Shared between Bogie and Axles)	
Centre of gravity, vertical	0.5335 m
Mass	3400 kg
Moment of inertia, roll	2134 kg·m²
Moment of inertia, pitch	1432 kg·m²
Moment of inertia, yaw	3506 kg·m²
Secondary Suspension	
Vertical stiffness (side springs)	1000 kN/m
Vertical stiffness (middle springs)	20000 kN·s/m
Longitudinal and lateral stiffness (side springs)	157 kN/m
Longitudinal and lateral stiffness (middle springs)	314 kN·s/m
Primary Suspension	
Lateral position	1.078 m
Vertical stiffness	782 kN/m
Vertical damper	60 kN·s/m
Longitudinal stiffness	53000 kN/m
Lateral stiffness	8000 kN/m
Other Dimensions	
Wheel spacing	1.85 m
Bogie spacing	14.19 m

about the longitudinal direction. For the connections between the track and the ground, the following coupling elements are used:

- Two vertical coil spring elements having lateral and vertical stiffness;
- Two vertical dampers with series flexibility and one lateral damper with series flexibility.

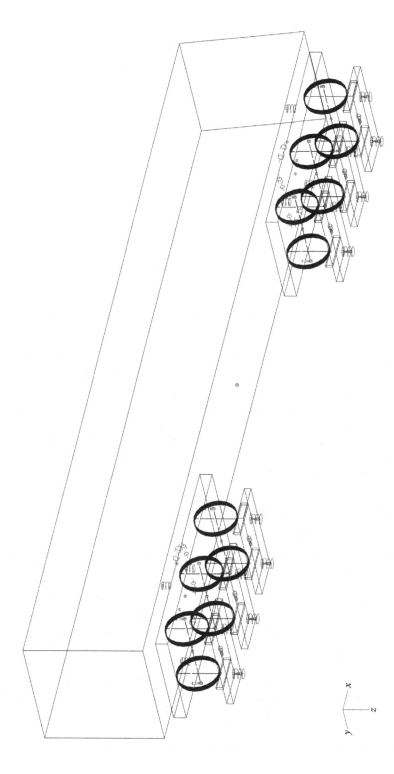

FIGURE 9.18 The model in Gensys multibody code.

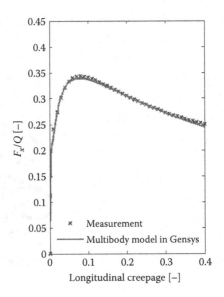

FIGURE 9.19   Gensys simulation results for the leading wheelset of the locomotive.

A simplified approach for the emulation of the traction control system has been introduced in the model, which allows smoothly changing the angular velocity of the wheelsets in order to reproduce the longitudinal creepage in the range from 0 to 0.4. The simulation results in the time domain for the leading wheelset of the locomotive using Gensys multibody software are shown in Figure 9.19.

Based on these results, it is possible to see that the behaviour of the contact model is stable and this allows its further application for numerical investigation. The initial task has therefore been completed.

9.4   SCENARIO C: REAL-TIME SIMULATION

Real-time models are one of the main resources required for the building of real-time computer systems, which can reproduce behaviour of rail vehicles or their component parts. These systems generally allow avoiding field tests by means of their replacement with laboratory experiments. Such an approach can provide significant benefits in the optimised development of new design solutions as well as a reduction in the costs of experiments because, in most field tests, it is quite difficult or almost impossible to reach critical or limit conditions due to the high risks involved. Costs connected with damage to rail vehicle and track infrastructure equipment, or potentially in the worse case, the loss of a human life or serious injuries, all make a combination of laboratory experiments and real-time modelling a much more acceptable approach.

In recent years, a great number of technical papers and manuscripts have been published in this field [27–36]. Some of the works discuss the challenges of the development of real-time contact models [28,29] or vehicle/bogie models [27,30,33] as well as test rig applications [28,31,34] or component testing [35]. The papers consider many existing challenges in this field based on the software-in-the-loop, processor-in-the-loop and hardware-in-the-loop simulation approaches (commonly referred to as SILS, PILS and HILS, respectively). All use real-time simulation models. However, HILS has found wide application in recent years [37] for the investigation of the behaviour of rail vehicles. This approach has great potential for further development in terms of providing solutions for the problematic aspects of rail vehicle component characterisation such as friction elements, pantograph systems, wagon connection elements and so on. The concept of HILS is that the problematic elements are tested physically in a laboratory rather than on an operational railway track, providing outputs to and receiving real-time inputs from the simulation system.

The HILS approach is based on the integration of different components which are parts of a complex mechatronic system:

- The hardware component or assembly under test (examples might be traction motors, suspension elements, bogies, etc.);
- A real-time simulation system equipped with real-time models of physical systems such as a train, a locomotive, or a wagon;
- Electronic control systems, equipment or units;
- Communications equipment and interfaces.

The technology provides great potential for rail vehicle development and testing in laboratories. The hardware component under study is set up in a laboratory for physical testing with force or torque actuators (motors can also be considered as torque actuators) attached at each point where the component would connect into the physical system. The physical system is usually simulated by specialised software developed to replicate the rail vehicle dynamics. In such software products, the developed mathematical or multibody model executed on real-time computer systems should run faster than real time. This issue for different varieties of application is well described in paper [38]. In our case, except for the electrical equipment, the main components of rail vehicles are mechanical systems with slow dynamics which require a simulation time step between 1 and 10 ms and, in some cases when it needs to describe a friction problem in an accurate way, the simulation time step might be significantly lower (from 100 to 500 μs) [38]. However, the commonly used time step for intercommunication processes is 1 ms [39]. Previously, most real-time simulations required the manual development of real-time models for mechanical systems. However, a recent publication shows that the application of multibody models is possible although verification of such models is required prior to their application [27].

In this section, the research studies published in [27] are going to be extended to a more complex simulation level where the calculation time response for the dynamic behaviour of the mechanical system (full-scale locomotive bogie test rig) under the traction mode will be considered.

Figure 9.18 shows the locomotive bogie test rig model as published in [27]. However, some modifications have been implemented in the model in order to represent its working under traction conditions as described in Section 9.3.

The model consists of a half locomotive car body, one bogie frame, six axle boxes, three wheelsets and six rollers. All bodies are modelled as rigid masses with

TABLE 9.10
Constraints on Bodies

	x-Longitudinal	y-Lateral	z-Vertical	f-Roll	k-Pitch	p-Yaw
Half-locomotive car body	$\sqrt{}$, $x = 0$	$\sqrt{}$	$\sqrt{}$	$\sqrt{}$, $f = 0$	$\sqrt{}$, $k = 0$	$\sqrt{}$
Bogie frame	$\sqrt{}$	$\sqrt{}$	$\sqrt{}$	$\sqrt{}$	$\sqrt{}$	$\sqrt{}$
Wheelset	$\sqrt{}$	$\sqrt{}$	$\sqrt{}$	$\sqrt{}$	$\sqrt{}$, $k = 0$	$\sqrt{}$
Roller	e	e	e	e	$\sqrt{}$	e

Note: $\sqrt{}$–Degree considered; e–degree eliminated; $x = 0$, $f = 0$ and $k = 0$ refer to longitudinal translation displacement plus roll and pitch rotations being fixed to be equal to zero.

six degrees of freedom. In order to reduce the calculation time and to obtain more stable behaviour of the mechanical system, some constraints have been set on these bodies (see Table 9.10). The basic parameters for the bogie test rig model, shown in Figure 9.20, are given in Table 9.11.

The connections (the secondary suspensions) between the locomotive body and one bogie frame include:

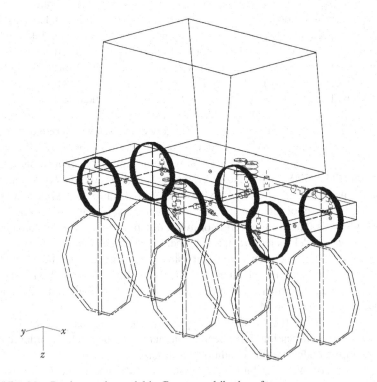

FIGURE 9.20   Bogie test rig model in Gensys multibody software.

- Four coil-spring elements in the vertical direction, each one including longitudinal, lateral and vertical stiffness;
- One stiffness element and one viscous damping element with series flexibility for the traction rod in the longitudinal direction specified by the coupling's attachment points (the viscous damper is provided to model the cushioning effect of the link bushing);

TABLE 9.11
Parameters for the Bogie Test Rig Model

Parameter	Value
Wheel spacing	2.38 m
Half-Locomotive Car Body	
Centre of gravity, vertical	1.930 m
Mass	87180 kg
Moment of inertia, roll	168550 kg·m²
Moment of inertia, pitch	3610410 kg·m²
Moment of inertia, yaw	3590650 kg·m²
Bogie Frame	
Centre of gravity, vertical	0.733 m
Mass	14860 kg
Moment of inertia, roll	6520 kg·m²
Moment of inertia, pitch	45370 kg·m²
Moment of inertia, yaw	50300 kg·m²
Axles (The Traction Motor Mass Shared between Bogie and Axles)	
Centre of gravity, vertical	0.565 m
Mass	2850 kg
Moment of inertia, roll	1789 kg·m²
Moment of inertia, pitch	1200 kg·m²
Moment of inertia, yaw	1789 kg·m²
Secondary Suspension	
Vertical stiffness	1069 kN/m
Vertical damper	40 kN·s/m
Longitudinal traction rod stiffness	25000 kN/m
Longitudinal damper	100 kN·s/m
Primary Suspension	
Vertical stiffness	730 kN·s/m
Vertical damper	5 kN·s/m
Longitudinal stiffness	3000 kN/m
Lateral stiffness	12000 kN/m
Lateral damper	15 kN·s/m
Longitudinal damper	560 kN·s/m
Longitudinal traction rod stiffness	25000 kN/m

- The secondary suspension has non-linear spring elements in the vertical and lateral directions, viscous vertical, lateral and yaw dampers (one lateral and two vertical bumpstops, two vertical viscous dampers, two lateral viscous dampers and two yaw dampers with series flexibility in the directions specified by the coupling's attachment points, respectively). No non-linear friction damping is modelled.

The wheelset is also modelled as a single mass with six degrees of freedom, including the pitch rotation by adding the wheelset's pitch moment. The connections (the primary suspensions) between one bogie frame and three wheelsets include:

- Six three-dimensional stiffness and damping elements in the X, Y and Z directions;
- Three lateral and six vertical bumpstops, and six vertical viscous dampers in the direction specified by the coupling's attachment points.

For the locomotive model with bogies with wheelset yaw relaxation, besides all the same connections as those in the locomotive model with rigid bogies, the additional connections in the primary suspensions are four viscous damping elements with series flexibility connected between the bogie frame and the first wheelset, and between the bogie frame and the third wheelset in the longitudinal direction specified by the coupling's attachment points.

Another significant issue in this simulation process is the implementation of wheel–roller contact in the model. When modelling the contact surface between wheel and roller in a multibody dynamics program, it is very important to take the radius of the rollers into account. When the vehicle is running along the track, the vertical curves of the track are so big compared to the wheel radius that the small curvature of the rails often can be neglected. However, in a roller rig, the rolling radius of the rollers is very close in size compared to that of the wheels, and hence the vertical curvature of the rails must always be taken into consideration. The rollers make the contact surface shorter in the longitudinal direction, which leads to a lower elliptical contact patch a/b-ratio and also a smaller size of the contact surface, which in turn leads to other creep forces in the contact surface. The standard contact coupling 'creep_polach2' based on the Polach theory while taking into account the issues mentioned above has been used in Gensys in order to implement the traction mode as stated in our aim for this section.

The integrator using the two-step Runge–Kutta (Heun) method has been chosen for simulation. This integrator is slightly slower than e1, but it provides more accurate solutions [27,40].

The model has been verified in the same way as published in [27] and no anomalies have been found in the model behaviour.

The calculation time has been estimated with the special time estimator realised in Gensys:

$$t_{tout} = t_{lsys} + t_{coupl} + t_{func} + t_{mass} + t_{cnstr} + t_{integ} + t_{ds} \qquad (9.7)$$

where t_{lsys} is the computational time spent on the position definition of local coordinate systems regarding the global coordinate system, t_{coupl} is the computational time

TABLE 9.12
Model Parameters for Adjustment of the Polach Model

Parameter	Value
μ_S	0.41
A	0.35
B (s/m)	0.45
k_A	0.6
k_S	0.13

spent on commands for coupling elements (coupling elements are elements of various types that connect masses to each other; for example, a coupling element can be a coil-spring, a rubber bushing, a hydraulic damper, a bumpstop, a friction element and so on), t_{func} is the computational time required for calculation of defined functions in the model script, t_{mass} is the computational time spent on mass commands (a mass command creates an inertia in the model masses, for example, the rail vehicle body, bogie components and wheelsets), t_{cnstr} is the computational time spent on constraint commands, t_{integ} is the computational time required for calculation inside the numerical integrator and t_{ds} is the computational time required for output data storage.

To provide simulation in the real-time mode, the generic kernel of the Ubuntu 64-bit operating system has been replaced with the open source real-time kernel developed by OSADL [41]. A computer equipped with Intel Core i7-3770 @ 3.40 GHz with 8 GB of RAM has been used for all simulations with the Gensys multibody software.

At the first stage, it is necessary to estimate time results as it has been done in [27]. This allows checking the computational times with a new implemented contact model based on the Polach theory and the changed integrator. The Polach model parameters are presented in Table 9.12.

The simulation results are shown in Figure 9.21. These results confirm that the developed multibody model can be easily used for real-time simulations. The calculation time for one time step is significantly less than 1 ms. That means that the model fully satisfies the requirements for real-time mechanical system models published in [38]. However, this model works in an unchangeable environment and does not have any excitations. The excitation factors can have an effect on the model behaviour, so it is necessary to verify the model with varying running conditions.

For such verification, the model should be structured to allow for situations when the influence of a traction control system also needs to be considered. For our model, a simplified traction control system has been introduced based on the individual wheelset traction control strategy (one inverter and one motor per axle) as a subroutine in Gensys. The system uses a feedback control strategy as shown in Figure 9.22, where T_{ref} is the reference torque; T_{ref}* is the reference torque generated by the control system; T_{in} is the input motor torque; T_{wheels} is the traction torque applied to the wheelsets; ΔT is the torque reduction; ω is the angular velocity of a wheelset; ω_r is the angular velocity of rollers; s_{est} is the estimated longitudinal slip and s_{opt} is the optimal longitudinal slip.

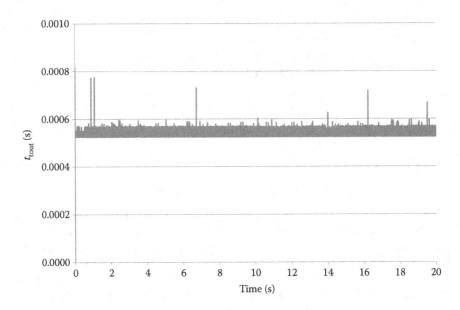

FIGURE 9.21   Calculation time results for real-time simulation in Gensys (integration and output time steps are both 1 ms).

The dynamics of inverters and traction motors have been modelled by means of the introduction of low-pass filters in the model. The torque limiter does not allow the control system to exceed the reference torque. The slip controller is a proportional-integral controller, which uses a slip error as the input signal to the controller.

The reference value for an optimal longitudinal slip (creepage) has been set equal to 0.08 for the dry friction condition.

The simulation results shown in Figures 9.23 through 9.25 confirm that the model's behaviour is stable under traction control conditions. The traction control system

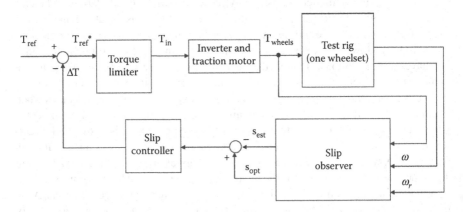

FIGURE 9.22   The traction control system for one bogie.

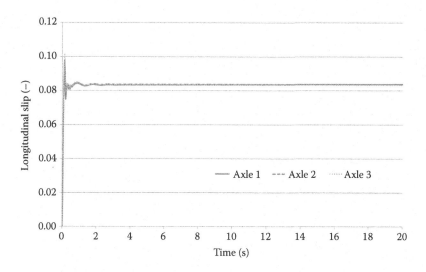

FIGURE 9.23 Longitudinal slips for real-time simulation with the implemented traction control system in Gensys (integration and output time steps are both 1 ms).

allows reaching the required values of the longitudinal slip and the traction coefficient. Small differences between a reference value of the longitudinal slip (creepage) and an estimated value can be easily explained by two reasons:

- A simplified traction control system based on the proportional-integral has been used in the model;

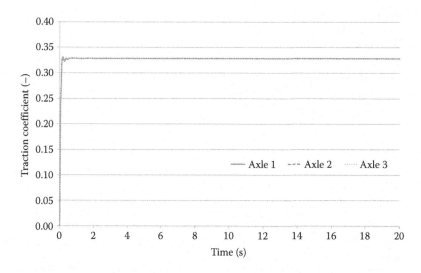

FIGURE 9.24 Traction coefficients for real-time simulation with the implemented traction control system in Gensys (integration and output time steps are both 1 ms).

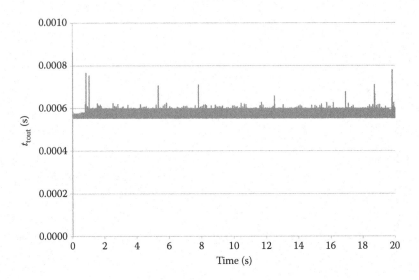

FIGURE 9.25   Calculation time results for real-time simulation with the implemented traction control system in Gensys (integration and output time steps are both 1 ms).

- A linear approach has been used for the calculation of the estimated value of longitudinal slip. This approach can provide some differences in slip (creepage) calculation in comparison with the non-linear approach used in Gensys. Some discussions about differences in both approaches can be found in [42].

The spikes in calculation time results are connected with the necessity to store calculation results on the hard drive for each time step. Such a problem can be avoided by means of an application of advanced real-time computer systems with a real-time network interface.

Summarising all simulation works performed in this section, it is possible to see that such a simulation technique can be easily adapted to any existing multibody software products and is useful at the initial stage of the development of real-time models for SILS, HILS and PILS.

REFERENCES

1. C. Cole, Longitudinal train dynamics, in *Handbook of Railway Vehicle Dynamics*, S. Iwnicki (Ed.), Taylor & Francis, Boca Raton, FL, 2006, pp. 239–277.
2. Centre for Railway Engineering, *CRE-LTS Reference Manual*, Central Queensland University, Rockhampton, Australia, 2001.
3. Federal Railroad Administration, Track Safety Standards Classes 1 through 5, *Track and Rail and Infrastructure Integrity Compliance Manual*, Vol. II, Chapter 1, Federal Railroad Administration, Washington, DC, July 2012.
4. O. Polach, M. Berg, S.D. Iwnicki, Simulation, in *Handbook of Railway Vehicle Dynamics*, S. Iwnicki (Ed.), Taylor & Francis, Boca Raton, FL, 2006, pp. 359–421.

5. M. Spiryagin, A. George, S. Ahmad, K. Rathakrishnan, Y. Sun, C. Cole, Wagon model acceptance procedure using Australian Standards, *Proceedings of Conference on Railway Engineering*, Brisbane, Australia, 10–12 September, 2012, pp. 343–350.
6. C. Matthews, Development of the Cv43ACi locomotive, *Proceedings of Conference on Railway Engineering*, Perth, Australia, 7–10 September, 2008, pp. 489–494.
7. N, Ramsey, F. Szanto, P. Hewison, Introducing the next generation locomotive to the Australian rail network, *Proceedings of Conference on Railway Engineering*, Perth, Australia, 7–10 September, 2008, pp. 471–480.
8. C. Cole, M. McClanachan, M. Spiryagin, Y.Q. Sun, Wagon instability in long trains, *Vehicle System Dynamics*, 50(Suppl), 2012, 303–317.
9. M. McClanachan, Y. Handoko, M. Dhanasekar, D. Skerman, J. Davey, Modelling freight wagon dynamics, *Vehicle System Dynamics*, 41(Supp), 2004, 438–447.
10. Y.Q. Sun, C. Cole, Comprehensive wagon-track modelling for simulation of three-piece bogie suspension dynamics, *Rail and Rapid Transit*, 221(8), 2008, 905–917.
11. Y.Q. Sun, S. Simson, Wagon-track modelling and parametric study on rail corrugation initiation due to wheel stick-slip process on curved track, *Wear*, 265(9–10), 2008, 1193–1201.
12. Y.Q. Sun, S. Simson, A nonlinear three-dimensional wagon-track model for the investigation of rail corrugation initiation on curved track. *Vehicle System Dynamics*, 45(2), 2007, 113–132.
13. Y. Sun, C. Cole, Vertical dynamic behaviour of three-piece bogie suspensions with two types of friction wedge, *Journal of Multibody System Dynamics*, 19(4), 2008, 365–382.
14. Y.Q. Sun, M. Spiryagin, S. Simson, C. Cole, D. Kreiser, Adequacy of modelling of friction wedge suspension in three-piece bogies, *Proceedings of 22nd IAVSD Symposium on Dynamics of Vehicles on Roads and Tracks (CD)*, Manchester, UK, 14–19 August, 2011.
15. Y.Q. Sun, C. Cole, Comprehensive wagon-track modelling for simulation of three-piece bogie suspension dynamics, *Journal of Mechanical Engineering Science*, 221(8), 2007, 905–917.
16. Y.Q. Sun, C. Cole, M. Spiryagin, S. Simson, Evaluation of heavy haul wagon dynamic behaviours based on acceleration measuring systems, *Proceedings of 10th International Heavy Haul Association Conference*, New Delhi, India, February 4–6, 2013, pp. 546–553.
17. J.J. Kalker, A fast algorithm for the simplified theory of rolling contact, *Vehicle System Dynamics*, 11(1), 1982, 1–13.
18. DEsolver, Users manual for program RUNF_INFO. See: http://www.gensys.se/doc_html/misc_runf_info.html.
19. DEsolver, Debugging a vehicle model. See: http://www.gensys.se/doc_html/analyse_check.html#Mainmenu.
20. L. Rawlings (Ed.), *VAMPIRE (Version 4.32) User Manual*, AEA Technology, Derby, UK, 2004.
21. DEsolver, Error in a vehicle model. See: http://www.gensys.se/doc_html/analyse_error.html.
22. Standards Australia and Rail Industry Safety & Standards Board, *AS 7509.2 Railway Rolling Stock–Dynamic Behaviour—Part 2: Freight Rolling Stock*, Sydney, Australia, 2009.
23. O. Polach, Creep forces in simulations of traction vehicles running on adhesion limit, *Wear*, 258, 2005, 992–1000.
24. O. Polach, Influence of locomotive tractive effort on the forces between wheel and rail, *Vehicle System Dynamics*, 35(Supp), 2001, 7–22.
25. M. Spiryagin, S. Simson, C. Cole, I. Persson, Co-simulation of a mechatronic system using Gensys and Simulink, *Vehicle System Dynamics*, 50(3), 2012, 495–507.

26. M. Spiryagin, Y.Q. Sun, C. Cole, S. Simson, I. Persson, Development of traction control for hauling locomotives, *Journal of System Design and Dynamics*, 5(6), 2011, 1214–1225.

27. M. Spiryagin, Y.Q. Sun, C. Cole, T. McSweeney, S. Simson, I. Persson, Development of a real-time bogie test rig model based on railway specialised multibody software, *Vehicle System Dynamics*, 51(2), 2013, 236–250.

28. N. Bosso, M. Spiryagin, A. Gugliotta, A. Soma, *Mechatronic Modelling of Real-time Wheel-rail Contact*, Springer, Berlin, Germany, 2013.

29. N. Bosso, A. Gugliotta, A. Somà, M. Spiryagin, Model of scaled test rig for real time applications, *ABCM Symposium Series in Mechatronics*, 5(VIII), 2012, 1288–1298.

30. Y. Maki, T. Shimomura, K. Sasaki, Building a railway vehicle model for hardware-in-the-loop simulation, *Quarterly Report of Railway Technical Research Institute*, Japan, 50(4), 2009, 193–198.

31. N. Watanabe, Y. Maki, T. Shimomura, K. Sasaki, T. Tohtake, H. Morishita, Hardware-in-the-loop simulation system for duplication of actual running conditions of a multiple-car train consist, *Quarterly Report of Railway Technical Research Institute*, Japan, 52(1), 2011, 1–6.

32. C.G. Kang, H.Y. Kim, M.S. Kim, B.C. Goo, Real-time simulations of a railroad brake system using a dSPACE board, *Proceedings of the ICROS-SICE International Joint Conference 2009*, Fukuoka, Japan, 18–21 August, 2009, pp. 4073–4078.

33. E. Meli, M. Malvezzi, S. Papini, L. Pugi, M. Rinchi, A. Rindi, A railway vehicle multibody model for real-time applications, *Vehicle System Dynamics*, 46(12), 2008, 1083–1105.

34. L. Pugi, M. Malvezzi, A. Tarasconi, A. Palazzolo, G. Cocci, M. Violani, HIL simulation of WSP systems on MI-6 test rig, *Vehicle System Dynamics*, 44(Supp1), 2006, 843–852.

35. I.V. Korzina, Imitation model of the electric locomotive for the test of microprocessor control systems, PhD thesis, Moscow State University of Railway Engineering (MIIT), 2006 (in Russian).

36. M. Spiryagin, V. Spiryagin, *Modelling of Mechatronic Systems of Running Gears for a Rail Vehicle*, East Ukrainian National University, Lugansk, Ukraine, 2010 (in Ukrainian).

37. M. Spiryagin, C. Cole, Hardware-in-the-loop simulations for railway research (Editorial), *Vehicle System Dynamics*, 51(4), 2013, 497–498.

38. J. Belanger, P. Venne, J.N. Paquin, The what, where and why of real-time simulation, OPAL-RT Technologies, October, 2010. See: http://www.opal-rt.com/technical-document/what-where-and-why-real-time-simulation.

39. M. Arnold, B. Burgermeister, C. Führer, G. Hippmann, G. Rill, Numerical methods in vehicle system dynamics: State of the art and current developments, *Vehicle System Dynamics*, 49(7), 2011,1159–1207.

40. H. Klee, R. Allen, *Simulation of Dynamic Systems with MATLAB and Simulink* (2nd ed.), CRC Press, Boca Raton, FL, 2011.

41. OSADL Project: Realtime Linux. *Open Source Automation Development Lab*. See: https://www.osadl.org/Realtime-Linux.projects-realtime-linux.0.html.

42. A.A. Shabana, K.E. Zaazaa, H. Sugiyama, *Railroad Vehicle Dynamics: A Computational Approach*, CRC Press, Boca Raton, FL, 2008.

10 Conclusion

The authors' intent in writing this book was to make available their knowledge of the highly complex technical subject of rail vehicle design and performance in various train configurations gained over diverse and lengthy careers in the rail industry, and more particularly in the field of railway research. General and advanced modelling techniques for both individual rail vehicle dynamics and longitudinal train dynamics are discussed. The text was structured so that basic issues and terminology have been covered before detailed explanations and techniques. Worked examples allow a virtual hands-on approach for those interested in actually carrying out simulations. We hope that readers have found the information flow easy to follow and understand. Many references are provided which will allow readers to further explore the international knowledge base that has developed from experience in operating various types of rail transport and, through research, to solve problems and improve train safety and performance.

Readers with enquiries regarding *Design and Simulation of Rail Vehicles* can contact the Centre for Railway Engineering at Central Queensland University by email at cre@cqu.edu.au or visit the website at www.cre.cqu.edu.au to find individual contact details.

Index

A

AAR, *see* Association of American
 Railroads (AAR)
ABAQUS FE rail modelling, 241
ABAQUS software package, 89
ABB, *see* ASEA Brown Boveri (ABB)
AC, *see* Alternating current (AC)
Acceptance tests, 24–25
ADAMS, *see* Automatic Dynamic Analysis of
 Mechanical Systems (ADAMS)
Adams/Controls plug-in, 253
ADAMS/Rail software, 120;
 see also Simulink
 rail simulations, 120–121
 for rail vehicle dynamics, 253–254
 rail vehicle engineers using, 121
Advanced simulation methodologies, 277;
 see also On-line simulation
 complex tasks and solutions, 277–278
 real-time simulation, 300–308
Aerodynamics, 88
Air brakes, *see* Pneumatic brakes
Air resistance, 191; *see also* Propulsion
 resistance
Air spring, 13, 55, 63
 elements, 111–112
 installation, 60
 secondary suspensions with, 111
 suspension, 58, 60
Alternating current (AC), 34
 electric traction for AC
  locomotive, 73, 74
 traction, 71–72
American Public Transport Association
 (APTA), 23
APTA, *see* American Public Transport
 Association (APTA)
ASEA Brown Boveri (ABB), 30
Association of American Railroads
 (AAR), 23
Augmented formulation, 106–107
Autocoupler, 21
 draft gear in, 22
 freight wagon, 21
Automatic Dynamic Analysis of Mechanical
 Systems (ADAMS), 120, 253
Autonomous rolling stock, 31
Axle box, 17, 53, 54, 59
Axle load, 32, 242

B

Ballast, 203, 215–216
 ballast–subballast pyramid
  submodel, 124, 214
 damping coefficients of, 215
 effects on vehicle dynamics, 92
 vibration of, 214
Bathtub, 10
BCP, *see* Brake cylinder pressure (BCP)
Beams on elastic foundation theory
 (BOEF theory), 203
Bearing box, *see* Axle box
Bearings, 15
 axle box, 17
 types, 17
 wheelset, 16
Bifurcation point, 231
Bilevel car design, 36
BOEF theory, *see* Beams on elastic foundation
 theory (BOEF theory)
Bogies, 14, 52
 centre-bowl, 14
 FEM analysis, 15
 heavy haul locomotive three-axle
  bogie, 54
 monomotor, 55
 passenger wagon, 14
 radial steering of wheelsets, 54
 rail vehicle bogie, 53
 strain instrumentation, 4
 three-piece bogie, 14
 three-piece freight bogie, 15
Bottom dump wagons, 9; *see also*
 Freight wagons
Bounce test, 25; *see also* Curving test;
 Roll test; Twist test
BPP, *see* Brake pipe pressure (BPP)
Brake cylinder pressure (BCP), 152
Brake cylinders, 18
Brake modelling, 84
 disc brakes, 88
 ECP brakes, 87
 pneumatic brakes, 85–86
 wheel brakes, 87
Brake pipe, 18
 fluid models use, 86
 models, 168, 169
 pneumatic brakes, 85, 87
Brake pipe pressure (BPP), 152

Printed in the United States
by Baker & Taylor Publisher Services